Modelling and Management of Engineering Processes

Peter Heisig · P. John Clarkson · Sandor Vajna
Editors

Modelling and Management of Engineering Processes

Springer

Dr. Peter Heisig
University of Cambridge
Cambridge Engineering Design Centre
Department of Engineering
Trumpington Street
Cambridge CB2 1PZ
UK

Prof. P. John Clarkson
University of Cambridge
Cambridge Engineering Design Centre
Department of Engineering
Trumpington Street
Cambridge CB2 1PZ
UK

Prof. Sandor Vajna
Otto-von-Guericke University Magdeburg
Faculty of Mechanical Engineering
POB 4120
39016 Magdeburg
Germany

ISBN 978-1-4471-5717-5 ISBN 978-1-84996-199-8 (eBook)
DOI 10.1007/978-1-84996-199-8
Springer London Dordrecht Heidelberg New York

British Library Cataloguing in Publication Data
A catalogue record for this book is available from the British Library

Softcover re-print of the Hardcover 1st edition 2010

Cover design: Peter Heisig, P. John Clarkson and Sandor Vajna

Printed on acid-free paper

Springer is part of Springer Science+Business Media (www.springer.com)

Preface

The importance of innovative processes for design and engineering in ensuring business success is increasingly recognised in today's competitive environment. However, academia and management need to gain a more profound understanding of these processes and develop better management approaches to exploit such business potential.

The aim of this *First International Conference on the Modelling and Management of Engineering Processes* is to showcase recent trends in the modelling and management of engineering processes, explore potential synergies between different modelling approaches, gather and discuss future challenges for the management of engineering processes and identify future research directions.

This inaugural Modelling and Management of Engineering Processes (MMEP) conference is being organised by the Engineering Design Centre at the University of Cambridge, the Chair of Integrated Product Development at the Royal Institute of Technology in Stockholm, the Institute for Product Development at the Technische Universität München, and the Chair for Information Technologies in Mechanical Engineering at the Otto-von-Guericke-Universität in Magdeburg on behalf of the Design Society's Special Interest Group of the same name.

This conference marks not only a new beginning for the MMEP community, but also the end of a process to develop a research roadmap for the Modelling and Management of Engineering Processes. During March 2009 a series of industry workshops were held in the UK, Sweden and Germany in order to identify future research needs, assisted by representatives from 27 companies from within the manufacturing, service and healthcare sectors. A preliminary roadmap was presented to and discussed with the research community in August 2009 at the ICED Conference in the US, and a joint white paper drafted (Heisig *et al.* 2009)[1].

---

[1] Heisig P, Clarkson PJ, Hemphälä J, Wadell C, Norell-Bergendahl M, Roelofsen J, Kreimeyer M, Lindemann U (2009) Challenges and future fields of research for modelling and management of engineering processes, 2nd edn. Workshop Report CUED/C-EDC/TR 148, Cambridge Engineering Design Centre, Department of Engineering, University of Cambridge, UK

This first MMEP conference is intended launch a bi-annual series providing an international platform to highlight and discuss industry best practice alongside leading edge academic research. A Programme Committee has been assembled, comprising representatives from fifteen countries including Australia, Argentina, Canada, France, Germany, India, Israel, Italy, Japan, Korea, the Netherlands, Spain, Sweden, USA and the United Kingdom. Industry interest is reflected in the fact that nearly half of the committee members represent global companies, such as Airbus, AUDI, BAE, BMW, Boeing, Bombardier, BP, BT, Daimler, General Motors, MAN, Infosys, Renault, Rolls-Royce, Samsung, SAP, Scania, Siemens, Toshiba, Volvo and Xerox.

The papers chosen for inclusion in this volume have been selected by reference to blind reviews undertaken by members of the Programme Committee who in turn represent a range of key disciplines including computer science, engineering design, innovation management and product development. They represent a sample of leading national and international research in the fields of engineering design, process modelling in engineering design and product development, and innovation management are divided into four areas:

I. *Engineering Process Management in Theory* concerns research that addresses process architecture frameworks, the nature of process modelling and product engineering processes;

II. *Managing Complex Engineering Processes* focuses on approaches to support the planning and improvement of engineering processes, highlighting issues such as uncertainty and iteration;

III. *Managing Product and Process Information* looks at research that supports the capture of information and knowledge, process mining and estimation methods based on sparse data;

IV. *Engineering Process Management in Practice* describes process management in organisations, including the modelling of process quality, interactive visualisation and robust innovation processes.

Finally, we would like to thank all those authors and reviewers who have contributed to the preparation of this book, and also Suzanne Williams, Susan Ball and Mari Huhtala who transformed a disparate set of initial drafts into a coherent and attractive book.

Peter Heisig, P. John Clarkson and Sandor Vanja
The MMEP 2010 Editorial Committee
July 2010

Contents

Part III Managing Product and Process Information

Part IV Engineering Process Management in Practice

List of Contributors

Albers A., Karlsruhe Institute of Technology, University of Karlsruhe, Karlsruhe, Germany
Braun A., Karlsruhe Institute of Technology, University of Karlsruhe, Karlsruhe, Germany
Buet G., Renault SAS, Boulogne-Billancourt, France
Chakrabarti A., Indian Institute of Science, Bangalore, India
Clarkson P.J., Engineering Design Centre, Department of Engineering, University of Cambridge, Cambridge, UK
Dumitrescu R., Heinz Nixdorf Insitute, University of Paderborn, Paderborn, Germany
Eckert C.M., Open University, Milton Keynes, UK
Fujita K., Osaka University, Osaka, Japan
Gausemeier J., Heinz Nixdorf Insitute, University of Paderborn, Paderborn, Germany
Gidel T., Renault SAS, Boulogne-Billancourt, France
Girard P., Université Bordeaux, Bordeaux, France
Gokula Vijaykumar A.V., Indian Institute of Science, Bangalore, India
Gonnet S.M., CIDISI-UTN/INGAR (CONICET-UTN), Santa Fe, Argentina
Grebici K., McGill University, Montreal, Quebec, Canada
Hahn A., OFFIS Institute for Information Technology, Germany
Häusler R., OFFIS Institute for Information Technology, Germany
Henning G., INTEC (CONICET-UNL), Santa Fe, Argentina
Hisarciklilar O., McGill University, Montreal, Quebec, Canada
Jiao R., Hong Kong Polytechnic University, Hong Kong, China
Kahl S., Heinz Nixdorf Insitute, University of Paderborn, Paderborn, Germany
Kerley W.P., Engineering Design Centre, Department of Engineering, University of Cambridge, Cambridge, UK

Kind C., Technical University of Berlin, Berlin, Germany
Koppe R., OFFIS Institute for Information Technology, Germany
Lari K., McGill University, Montreal, Quebec, Canada
Le H.N., Engineering Design Centre, Department of Engineering, University of Cambridge, Cambridge, UK
Leone H., CIDISI-UTN/INGAR (CONICET-UTN), Santa Fe, Argentina
Li C., Hong Kong Polytechnic University, Hong Kong, China
Lindemann U., Technical University of Munich, Munich, Germany
Liu Y., Hong Kong Polytechnic University, Hong Kong, China
Merlo C., ESTIA Innovation, Bidart, France
Millet D., Renault SAS, Boulogne-Billancourt, France
Muschik S., Karlsruhe Institute of Technology, University of Karlsruhe, Karlsruhe, Germany
Nomaguchi Y., Osaka University, Osaka, Japan
Paetzold K., Bundeswehr University Munich, Munich, Germany
Pol G., ESTIA Innovation, Bidart, France
Poppen F., OFFIS Institute for Information Technology, Germany
Reitmeier J., Bundeswehr University Munich, Munich, Germany
Robin V., Université Bordeaux, Bordeaux, France
Roelofsen J.M.K, Technical University of Munich, Munich, Germany
Roldán M.L., CIDISI-UTN/INGAR (CONICET-UTN), Santa Fe, Argentina
Stacey M.K., Faculty of Technology, De Montfort University, Leicester, UK
Stark R., Technical University of Berlin, Berlin, Germany
Suss S., McGill University, Montreal, Quebec, Canada
Thomson V., McGill University, Montreal, Quebec, Canada
Woll R., Technical University of Berlin, Berlin, Germany
Wynn D.C., Engineering Design Centre, Department of Engineering, University of Cambridge, Cambridge, UK
Yamamoto T., Osaka University, Osaka, Japan
Zhang H., Tsinghua University, Beijing, China

Part I

Engineering Process Management in Theory

Chapter 1

What is a Process Model? Reflections on the Epistemology of Design Process Models

C.M. Eckert and M.K. Stacey

1.1 Introduction

It is now established wisdom in the engineering design community that models are useful means of understanding and interacting with both products and processes. However this is a fairly recent development. A hundred years ago at the height of British engineering prowess, this was not the case. While it is possible that individual companies used informal models to describe their processes, formal prescriptive process models only emerged around World War Two and none of the current established process modelling techniques existed. After seventy years of experience, what can be expressed in process models, and how they can be used and interpreted, is still not fully understood. This paper describes some of the challenges that companies face in getting benefit out of process models and argues that some of them arise from the nature of models, as analysed by philosophers, and the nature of process models in particular.

Product models did exist. The earliest models of products had the same form as the larger thing to be created. As the complexity of the things made increased, product models became more abstract and selective. Engineers sketched, created schematic models of their products, and used graphical approaches to solve problems that are now solved numerically by a computer at the press of a button. Models serve many purposes in design, not just as visualisations of a product or process, aiding in idea generation, problem solving and evaluation, but also in facilitating the interaction of team members. How well a model can carry out these functions depends on the quality of the model.

However, assessing how closely a model describes its target (the object or process that it models) is far from simple. The relationship between a model and its target is fundamentally analogical, requiring the construction of a mapping between dissimilar things that have similar combinations of relationships between their parts (see Holyoak and Thagard, 1996). When the models are physical objects, the degree of abstraction required to see similarities between components and patterns of relationships is not that great, but for mathematical and schematic models, making a connection between equations and diagrams and the form and behaviour of physical objects requires a much greater imaginative leap. But the relationship between a process model and a process is much trickier and elusive still, not least because process is itself an elusive concept.

Complexities arise from process models mixing 'is like' with 'should be like' (Section 1.2), the subtleties of what exactly a model *means* (Section 1.3), and conflicts in how models can be constructed and interpreted (Section 1.4), as well as the wide variation in scope and detail afforded by different modelling approaches (Section 1.5). We discuss the different roles models play in industry and the problems in building and using models we have observed in industrial practice in Section 1.6, many of which are inherent in the nature of models; and relate our analysis of the conceptual complexities involved in process modelling to philosophical analyses of the role of models in science, a simpler problem that has attracted far more attention from philosophers than has design.

1.2 'Is' versus 'Should Be'

The term *process* is used in the design community to refer to two distinct but related concepts: (a) the sum of the actual activities that are or should be carried out to design a product or component or to solve a design problem; and (b) an abstract and relatively general conceptualisation of what is done or should be done to design a product, that is meant to apply to a class of products, an organisation, or an industry sector. While many different kinds of models are used to describe the process in the sense of what designers actually do, processes as abstract conceptualisations are expressed in idealised and abstract models. For both meanings of process, what is being modelled by a process model can differ and be open to debate. For models of detailed behaviour, there are three questions, whether the model is an accurate description, whether it is understood correctly, and whether the process is the right process. For processes as abstractions, the question is whether the model is a correct or appropriate conceptualisation for thinking about the structure of design activities.

Process models are employed for a number of distinct but interrelated purposes: to understand a process; to plan a process by determining what needs to be done when; to prescribe a procedure to be followed; and to predict how a process will behave, for example through simulations.

Process models divide, at first sight, into *descriptive* models - scholarly analyses of what happened in a process, or what happens in a category of processes - and *prescriptive* models - what designers *should* do to produce a particular type of product. For descriptive models, whether the model achieves its purpose is a question of whether it provides valid understanding. The relation between model and target has been extensively discussed by philosophers of science, but conformity depends essentially on the *accuracy* of the mapping between the structure of the model and the structure of the target. For prescriptive models, the relation between model and target is deontic, defining what designers should do: the model precedes its target, and reality conforms, or not, to the model.

This picture is complicated by models that capture best practice in frequently performed activities. These models attempt to describe how designing actually gets done, but only in selected or idealised cases, rather than as an accurate reflection of the range of behaviour in the category of activities they cover, and their purpose is to function as prescriptive models for future designing. Thus they have different conformity relationships to past and future processes, and the actual target of a description of best practice may not be clearly either an abstraction or an identifiable chunk of reality. In practice this distinction is blurred because in the attempt to generate descriptive models, one is confronted with a combination of descriptions of actual behaviour, as-is, and an account of what should have happened or should happen, as-should-have-been.

1.3 Philosophy of Science Perspective on Models

The philosophers of science have thought very hard about what a model is and what it can represent, though the central importance of models in science was recognised remarkably late. Carnap (1938) famously remarked that 'the discovery of a model has no more than an aesthetic or didactic or at best heuristic value, but it is not at all essential for a successful application of the physical theory'. This is echoed in an even more dismissive view of high level abstract models in design by Lawson (1980), cited in Wynn and Clarkson (2005), that they are '...about as much help in navigating a designer through his task as a diagram showing how to walk would be to a one year old child...Knowing that design consists of analysis, synthesis and evaluation will no more enable you to design than knowing the movements of breaststroke will prevent you from sinking in a swimming pool'.

However, since the 1960s models have been placed at the heart of the philosophical understanding of science by construing theories as families of models (Suppe, 1989). Models have become recognised as a vital part of acquiring and organising scientific knowledge, because they represent the target in some form. Models are seen as isomorphic or similar to the target, whereby philosophers such as Suppe (1977) see a model essentially as a representation of the structure of its target. However Frigg (2003) explains representations in terms of three relations: denotation, display and designation: 'A model denotes its target system in roughly the same way in which a name denotes its bearer. At the same time it displays certain aspects, that is, it possesses these aspects and a user of the model thematises them. Finally, an aspect of the model designates an aspect of the target if the former stands for the latter and a specification of how exactly the two relate is provided'.

The representation comprises two mappings: an abstraction of real or imagined physical or social phenomena to a conceptualisation of how they work, and an analogy between the structure of this conceptualisation and the structure of the mathematics or objects and connectors that comprise the model itself. These abstract conceptualisations parameterise or strip out the contingent details of individual cases, such as exactly which parts are listed on a bill of materials, or safety margins of components, to define a category of possible situations sharing essential characteristics but free to vary in other ways. More abstract conceptualisations cover broader classes of superficially dissimilar things. This implies that abstraction is a process of selecting certain features and relationships as important and thereby discarding other features. Abstraction is driven by a particular purpose rather than being an absolute property of the range of cases it covers. This selection of significant features is biased by what seems to be important at a particular time from a particular viewpoint. Abstraction is often generalisation over multiple similar objects or situations, whether concrete cases, idealised descriptions or skeletal memories, so that common features are selected to form the abstraction. Frigg (2003) argues that abstraction only exists if there is also one or a group of more concrete concepts that could provide a more specific description. A related dimension is the level of detail of a description or model.

Models in advanced sciences such as physics and biology are abstract objects constructed in conformity with appropriate general principles and specific conditions. Scientists use models to represent aspects of the world for various purposes, exploiting similarities between a model and that aspect of the world it is being used to represent, to predict and explain (Giere, 2004). Giere argued that models are not linguistic and do not have truth values; rather, that theories also comprise linguistic statements about the similarity relationships between model and reality, which do have truth values. However, while the mathematical models that are central to physics can be seen as clearly distinct from assertions about how they relate to the physical universe, separating the formal structure that makes a model a model from how it relates to its target is more problematic when the definition of the model takes the form of a linguistic description of its target. Nevertheless to constitute a model, the description defines an ideal type to which the targeted aspect of reality may or may not conform.

In the physical sciences, failure of a model to match observations is evidence for the inadequacy of a theory, though the conformity relationship is seldom seen as binary, and almost-successful theories are revised rather than abandoned (see Lakatos, 1970). Designing involves a complex interaction of cognitive and social phenomena. In the social sciences, models are more often conscious simplifications of complex situations that are partly dependent on contingent circumstances beyond the scope of the model, so the causal factors included in the model account for part of the similarities and differences between cases. Sometimes this can be quantified statistically in terms of the proportion of the variance in the values of the dependent variables that is accounted for by the independent variables included in the model. In these situations, it makes more sense to talk about how closely reality approximates to the model - degree of similarity.

However in design, process models are often prescriptive specifications, or are idealisations of best practice. Then the conformity relationship between a process and its model is rather more complex, because conformity is often not just a matter of 'is' or 'was', but includes deontic relationships: 'should' and 'must'. However, being accurate is not what a prescriptive model is there for, which primarily is enabling people to make inferences about how they should act to achieve a successful process. It might be more fruitful to speak in terms of appropriate models, that are fit for purpose, or inappropriate models, which fail to fulfil the requirements that are placed on them.

1.4 Interpreting Models: A Perspective from Social Science

Models are built for specific purposes by people, who select and represent the aspects of the process they consider to be important. They are understood and interpreted by different people according to their own goals, their own understanding of what they are doing and why they are doing it, and their own understanding of the social and organisational context they are working in. Each participant in a design process uses his or her distinct understanding of that process, informed by the models they use, as an information resource for guiding his or her actions. Process models serve as *boundary objects* (see Bucciarelli, 1994), whose labels for activities, types of information and communication channels convey information between specialists in different fields with different expertise, who see very different meaning in them.

The most well-developed and widely used approach to modifying and improving sociotechnical systems, *soft systems methodology* (Checkland, 1981), starts from the premise that the consequence of the subjectivity of individual understanding of social processes such as designing by the people participating in them is that they have *no* objectively true structure. Instead, any account such as a design process model constitutes one subjective viewpoint that isn't necessarily more true than someone else's contradictory view; any shared understanding is both partial and the outcome of a social process of negotiation. Hence, treating a social system as though it has an objectively true or correct structure is at best a pragmatically useful compromise - one that Checkland argued is misleading and should be avoided. We do not fully share Checkland's scepticism about as-if-objective descriptions of social structures and processes. But it is important to recognise that a very high degree of consensus about a skeleton of social facts does not imply a shared perception of their implications, and that alternative views of how processes work may be equally legitimate and valid accounts of how different people experience them.

Ethnographers have put forward the view that an ethnographic account of a social system is one possible valid partial truth among any number of true partial accounts, and the test of validity is that it looks right to the participants (see Hammersley and Atkinson, 1995). However for process models, it is not enough that the model looks right to all the people involved in generating it. It must also be useful to those who just pick it up. It must enable its users to make valid and valuable inferences about how the process behaves and how they should act.

Design process models are interpreted by their users - and how they are interpreted by the participants in the process can be significant for how designing happens. Design process models are interpreted differently by different people in two ways. First, people differ according to their perspectives, knowledge and experience in the range of possible concrete situations and actions they can imagine as possible, or regard as feasible or appropriate, hence in what range of possible processes they see a model as encompassing. Disagreements may stem from mismatches in assumptions rather from consciously articulated beliefs. Second, people differ in how they interpret the epistemological status of the model, as mandatory specification of actions, or of goals to be achieved, or as guidelines to be used or adapted as each new situation demands.

1.5 Models Vary in Scope and Detail

Rather than reviewing different modelling approaches, this section reflects on the categories of models and some of the conceptual challenges of modelling design processes. Wynn and Clarkson (2005) review generic process models; and Browning and Ramasesh (2007) review modelling techniques for specific processes. Regardless of the type of model used, models are affected by the following factors:

- **Selection**: Any model is a form of abstraction, as it requires people to select what they consider to be significant. This selection applies both to the features which are included and to the scope of the model. For example many process models do not explicitly include the documentation that designers produce, as they don't consider it a design activity; nevertheless it is a vital part of any design process, especially for certification.
- **Representational bias:** The type of process model determines what can be described in the model and therefore biases the selection of elements that are included in a model. For example stage gate processes do not include iteration (although in practice, failed stage gate reviews sometimes need to be repeated, so iteration can happen at that level of abstraction). Flowcharts make iterations explicit, but do not indicate location in time.
- **Modelling choices:** Even within one representation of formalism, the choices of the modeller determine what properties the model displays. For example if there is frequent iteration between two tasks, they can be shown as two tasks and an iteration loop or as just one task, in which case the iteration would be buried. If this iteration later causes problems, the simpler model would have hidden the behaviour, and therefore would not have been fit for purpose.

In many design domains models can be developed or proposed based on observations of multiple instances of similar phenomena. There are practical limitations to the number of cases and level of detail it is possible to look at. For example Eckert (2006) could look at multiple knitwear design processes, because they are short and quite simple, compared to engineering processes, like aircraft design, which require millions of person hours. While it is impossible to compare large-scale processes in their entirety, it is of course possible to look at tasks that are performed in a similar fashion in many processes. Therefore it is possible to have a set of generic models of limited scope, *e.g.* of stress analysis of certain components, but it is rather more difficult to produce a model of the process of, say, an entire jet engine. The variation between projects lies in the way these various repeatable processes are combined. Few people have the required overview to see similarity across multiple projects (see Flanagan *et al*, 2007).

Detailed models describe repeatable activities and are quite easy to produce, *e.g.* how to do stress analysis. High level models of the entire process in terms of a few selected core activities also remain fairly constant, *e.g.* each car requires styling, design of the engine and power train, design of the interior *etc.* Very large organisations can have a cascade of models - broad-brush stage gate processes for the whole company, and more detailed adaptation of these processes for local conditions and products that are required to conform to the models enshrining company policy.

The difficult modelling problem is the middle ground of models with a broad scope and a meaningful level of detail. Connectivity models such as design structure matrices (DSMs) try to capture dependencies between the elements of a model, but are notoriously difficult to build. Specific process models for process planning are therefore a special and difficult case. Not only can they not be based on multiple instances, they cannot be based on any instances at all that they actually apply to. They are the result of the application of abstract knowledge and reasoning by analogy from past experience.

1.6 Process Modelling in Industrial Practice

The empirical insights reported here are based on various case studies that were concerned with modelling and planning design processes in complex engineering domains.

In 2000 an empirical study was carried out in an automotive company to understand the range of models that designers used to plan design processes. Over a period of six weeks 18 experts were interviewed in one company on two sites, including engineers, engineering managers and business managers. This revealed many of the divergent views on models and applications for models (see Eckert and Clarkson, in press). In 2004 to 2005 we were also involved in studies of processes in another automotive sector company, which had worked hard on introducing Six Sigma and other prescriptive approaches across the organisation (see Flanagan, 2006).

The most recent process modelling activity we were involved in took place from 2005 to 2008 in an aerospace company where the early conceptual design process was modelled in P3 with the view of developing a new process for conceptual design (Eckert *et al.,* 2008).

All studies consisted of a combination of interviews, which were recorded and largely transcribed and informal conversations with designers and design managers. Many of the attitudes reflected here emerged as throwaway comments during interviews and were then taken up in informal conversations with design managers.

1.6.1 Models Serving as Process Plans

Rather than just employing process models to plan design processes, designers and design managers make use of different models for product, process and quality. Product models such as cost models or bills of materials are used to track progress through the process. Process models are generated - often informally - by individuals and teams, and on a formal level for the entire project drawing both on activities and prescribed stage gates. However as different parts of a design don't evolve at the same rate, lead time plans and test schedules are developed at a component level. To cope with emerging crises teams often develop fire fighting schedules, which deal with one issue to the exclusion of others, thus often propagating the problem. Eckert and Clarkson (in press) found that rather than have an explicit procedure to co-ordinate these different models and resolve conflicts between them, key stakeholders interact with more than one plan and bring these together as and when required.

Single integrated models are unlikely to exist; projects are modelled in a combination of different models. How these models relate to each other is often not made explicit. For example, for a flowchart style model, individual tasks are often broken down into further flowcharts. For the lower level models to make sense in their own right they need to repeat elements of the higher level models, without clearly indicating this. This duplication issue is most critical for time buffers, where everybody would like to add their own buffers, but not everybody needs to add buffers for the same iterations or unexpected tasks.

In the case study companies designers or design managers were given little advice on how to generate their process models or explicit instructions on what to include. While it is relatively obvious what needs to be in a product model, it is less clear what needs to be included in an activity plan, either the level of detail required or the type of activities that are included. Therefore the responsible people based their plans on what they were familiar with, not just in the style of the plans, but also in content. The engine company had developed a sequence of similar products by incremental improvement, and had a number of very experienced engineers who were able to mediate between different teams (see Flanagan *et al.*, 2007), so had the knowledge needed to develop process plans. The situation was more challenging in the car company, which tried to design a new car with a largely newly assembled team at a second site. As many people had recently joined the organisation, they developed models of very different styles and content, based on the models they had seen in their old teams.

1.6.2 Challenges with Building Models in Industry

All these models are partial in two ways: (a) By the way they are constructed they focus on particular aspects of the process, such as the sequence and flow of documents, or testing schedules. Being focused is an inevitable property of models. (b) They are also limited in their scope. Both automotive companies had systematic integrated product models, such as bills of materials, which also had a process function, but the process models only covered parts of the process.

The product and the process constrain each other. If process models are generated before the process has unfolded, the models are based on a prediction of what the product will look like and therefore what process it requires. If the process is determined in advance, it limits the range of possibilities for the product.

Typically the engineers try to relate what they know they have to do to what they had done previously on other projects, and reason by analogy to adjust the process that they know accordingly. If they think it will require significant new work, they might add more time to individual tasks and add tasks for validation. If the design is expected to be routine, they might reduce the expected time. As everybody has different experiences process models therefore are to some extent subjective in the way they are constructed.

Process models are also affected by the personality of the person who builds it. While some people are very cautious and base their process models on the worst case scenario, others are optimistic and provide models based on an ideal path. This is most pertinent in time estimates, but can also affect the structure of a model. It is not possible to see from the model itself whether an estimate is based on the best case, the worst case, the most typical case, or the most recent case. Even if directives were given for that or the creators of models would offer an explanation, this could not be taken for granted.

Organisational culture can discourage accurate planning or modelling of processes. If the way people book their hours is ill-aligned to the plans that they generate, then the record of the process will be a mismatch with the plan for the project, just because people needed to book time against the tasks of the particular project they mainly work on. Providing accurate plans for processes is often not rewarded in organisations, who praise people for process plans that are ambitious, rather than plausible. If over-ambitious plans fail, the people are often rewarded again for finding a way to rescue the situation. This nurtures a fire fighting culture, whereby teams concentrate on sorting out individual problems, often to the detriment of other tasks going in parallel.

1.6.3 Attitudes to Processes and Process Models

This section reflects about the attitudes to process models that engineers convey in the interviews we have conducted. This is an interpretation of throwaway comments of interviewees and complaints of project managers. However, in discussions with senior engineering managers, these observations rang true to them.

While it is possible to identify some of the ways in which models are wrong, by being incomplete, defining impossible or implausible task sequences, or making statements that are factually wrong, it is inherently difficult to say whether a model was "right" or "fit for purpose". People talked about "right" and "wrong" process models in very similar terms to product models, where comparisons can often be made between product and model. This lack of reflection about what a model can provide lies at the heart of the attitudes to process models discussed here.

The way models are constructed and their inherent properties make it very difficult to interpret how good a model is and what it can give people. None of the engineers we discussed models with are much aware of the practical and conceptual difficulties of creating, using and introducing models. They did not comment on the inherent limitations of modelling and seemed not to be aware of the practical limitations of models. They also showed no awareness that other people might interpret the models in a very different way. They assumed that their understanding of the model was "the" sensible interpretation and therefore assumed others would reach it as well.

However the resulting interpretations were very different. Some people saw process models as "schedules" that they could follow and needed to follow. Like a "timetable" showing the sequence of stations, the plan shows the sequence of activities. If things don't work out, they are disappointed by the people who carry out the process. The other end of the spectrum are people who see processes and process models as vague indications of what they might have to do, rather than anything binding, because they sense that something that is that "wrong" cannot be prescriptive.

People sometimes dismiss process models if some aspects of them are wrong, such as duration or order of tasks, rather than recognise that the models could still be useful if some of the information is wrong or no longer relevant, in the same way that a train timetable is still useful when the train is late, because it shows the route the train travels and the journey time on the plan is at least the minimum it can take.

Process champions, that is, people who really push a particular process, model or modelling approach, are very important in an organisation. Without them introducing a process is very difficult and people would not provide models of their individual activities. However, too much zeal can be very off-putting. If the champion pushes too hard people believe that rather than supporting them in successfully bringing the project to a conclusion, they are only interested in making sure their model is adopted. They end up following a process because they have to rather than because they see merit in it.

1.7 Conclusions: Applying Process Models

This paper has argued that in practice creating, handling and using process models is still problematic. While engineering activities are recurrent at the level of local tasks, where the procedures can be modelled based on multiple projects involving the same activity, and involve the same key tasks and major phases at a high level, they are very difficult to model at a medium level of abstraction, where people do not have the required overview and cannot draw on experiences with multiple similar products, because the specifics of projects were different.

Some of the problems are caused by management issues. However, many problems arise from the nature of models themselves. Many designers and managers do not understand what information a process model can provide them with and think of them in a similar way to product models.

It is important to recognise the properties of process models in managing and directing a process, because process models constrain and shape the processes to which they are applied, through the same aspects that make them useful. They provide information about the structure, dependency and preferred order of the activities and often contain rich additional information in terms of resources, constraints, times and so on.

A process model does not need to be totally correct in order to be useful, nor does correctness guarantee usefulness. The coarser the models, the more likely they are to be correct, but the less assistance they provide. A model that is wrong in some respects can be useful, as long as the users understand what the model can provide.

Using models to understand, develop and follow design processes can be problematic and lead to confusion and misunderstanding at four levels. First, a process might simply be misunderstood, or a prescriptive process might be an inappropriate way to achieve its intended result. Second, a process may be conceptualised, validly, in different ways according to the perspectives and priorities of different people. Third, as modelling involves prioritisation and selection, a shared understanding of a process may be modelled in different ways. Fourth, a model is interpreted by its different users according to their own experiences, knowledge and concerns, which may include conflicting unarticulated assumptions. All of these are significant issues in industrial practice.

Process models are built by individuals based on their own experiences for a particular purpose, which is usually not stated explicitly. Modelling inherently involves selecting some aspects and excluding others. One could argue that process models are always wrong because models exclude factors which might become important, because of the fact that they are excluded from the model. A process model draws the attention of the user to particular aspects of the process, tempting them to ignore others. If a model shows up potential problems, such as resource shortages in critical areas or the lack of interaction between important tasks, this encourages people to try to fix these problems, so that the problems go away. In a prevailing fire fighting culture this is likely to open up a gap somewhere else.

Rather than being a depiction or a set of instructions a process model can be thought of as a self-fulfilling prophecy. A prescriptive process model, or, more subtly, any model used to derive process information during designing, makes the process behave in the way it describes, because it is used to manage and understand the process for which it exists. Models here set goals and people are assessed against them. However they also provide a context in which isolated tasks or activities are carried out, and therefore shape these activities towards the expectations they generate about the rest of the process. Descriptive process models can also exert an influence on behaviour, because as they become the record of what has been, they influence the memory of the participants, and influence other people's understanding of how to design. As process models are inherently open to different interpretations, these interpretations of what the structure of the process is and how it is meant to guide behaviour influences what the process turns out to be. These interpretations may be in conflict.

It is important that designers and managers models can play in an organisation needs to be negotiated within a team and an organisation. An understanding of a model is a cognitive construct rather than an inherent property of the model, and a shared understanding is constructed through social processes of discussion and clarification understand that models can be interpreted in different ways and that people have different expectations of them. Rather than prescribing an interpretation of models, the understanding of the role that models can play in an organisation needs to be negotiated within a team and an organisation. An understanding of a model is a cognitive construct rather than an inherent property of the model, and a shared understanding is constructed through social processes of discussion and clarification.

1.8 Acknowledgements

This research has benefited from many conversations with Claudia Eckert's former colleagues at the Cambridge University Engineering Design Centre, and the EDC's financial support by the EPSRC. We are grateful for the assistance of our industrial collaborators, especially the engineers and managers we interviewed and observed. We thank our reviewers for their constructive and helpful comments.

1.9 References

Browning TR, Ramaseh RV (2007) A survey of activity network-based process models for managing product development projects. Production and Operations Management 16(2): 217-240

Bucciarelli LL (1994) Designing engineers. MIT Press, Cambridge, MA, US

Carnap R (1938) Foundations of logic and mathematics. In: Neurath O, Morris C, Carnap R (eds.) International Encyclopedia of Unified Science, University of Chicago Press, Chicago, IL, US

Checkland P (1981) Systems thinking, systems practice. Wiley, Chichester, UK

Eckert CM (2006) Generic and specific process models: Lessons from modelling the knitwear design process. In: Proceedings of the 6th International Symposium on Tools and Methods of Competitive Engineering (TMCE 2006), Ljubljana, Slovenia

Eckert CM, Clarkson PJ (in press) Planning development processes for complex products. Research in Engineering Design. Available at: http://www.springerlink.com/content/nr87165gl6r06223/fulltext.html (Accessed on 1 December 2009)

Eckert CM, Kerley WP, Clarkson PJ, Moss M (2008) Design for service: The new challenge for long-life products. In: Proceedings of the 7th International Symposium on Tools and Methods of Competitive Engineering (TMCE 2008), Kusadasi, Turkey

Flanagan T (2006) Supporting design planning through process model simulation. Cambridge Engineering Design Centre, University of Cambridge, Cambridge, UK

Flanagan T, Eckert CM, Clarkson PJ (2007) Externalising tacit overview knowledge: A model-based approach to supporting design teams. Artificial Intelligence in Engineering Design, Analysis and Manufacturing 21: 227-242

Frigg R (2003) Re-representing scientific representation. PhD thesis, Department of Philosophy Logic and Scientific Method, London School of Economics, London, UK

Giere RN (2004) How models are used to represent reality. Philosophy of Science 71: 742-752

Hammersley M, Atkinson P (1995) Ethnography: Principles in practice. Routledge, London, UK

Holyoak KJ, Thagard P (1996) Mental leaps. MIT Press, Cambridge, MA, US

Lakatos I (1970) Falsification and the methodology of scientific research programmes. In: Lakatos I, Musgrave A (eds.) Criticism and the growth of knowledge. Cambridge University Press, Cambridge, UK, pp 91-196

Lawson BR (1980) How designers think. Architectural Press, London, UK

Suppe F (1977) The search for philosophic understanding of scientific theories. In: Suppe F (ed.) The structure of scientific theories, 2nd ed. University of Illinois Press, IL, US

Suppe F (1989) The semantic conception of theories and scientific realism. University of Illinois Press, IL, US

Wynn DC, Clarkson PJ (2005) Models of designing. In: Clarkson PJ, Eckert CM (eds.) Design process improvement - a review of current practice. Springer, London, UK, pp 34-59

Chapter 2

Uniqueness and the Multiple Fractal Character of Product Engineering Processes

A. Albers, A. Braun and S. Muschik

2.1 Introduction

Every leaf of a tree has a different shape. One cannot understand form and function of a leaf without considering it as part of a whole tree with connections to the soil, the climate in which it grows and the animals living with it. One cannot succeed in describing a whole system from the point of view of one single part of it.

(Sandmeyer, 2009)

This metaphor, taken from an article about the Christian church, is also valid for the complex system of product engineering. Every product engineering process is unique and individual. This is the first out of five hypotheses about product engineering processes (Albers, 2010). In this paper we investigate where the differences between product engineering processes originate from. We examine the integrated product engineering model (iPeM) and its subsystems - the triple systems: *system of objectives*, *system of objects* and *operation system* (see Section 2.3) - and point out reasons for distinctions between individual processes. Not only processes themselves are individual but also their sub-processes with their networked hierarchies of activities, hierarchic levels of systems of objectives *etc.*

In Section 2.2 we review the iPeM as a model of engineering processes, its different abstraction levels and the related state of the art. The sections thereafter analyse the operation system, the system of objectives and the system of objects. It is shown that each of the systems is fractal and can be modelled self-similarly on different hierarchical levels in the iPeM. Examining the interconnections of the subsystems on different levels, we elaborate how different boundary conditions and restrictions in the system of objectives lead to individual subsystems and cause individual and unique engineering processes. The individual subsystems of the different fractal levels are interconnected in various ways and across hierarchic levels. Iterations or unexpected changes of objectives which require redesign have an especially strong impact on the engineering process and cause complexity which needs to be handled by contemplable approaches.

In Section 2.4 we present two examples for a wiki-based implementation of the approach. Both applications are meant to support product engineering processes by modelling and keeping track of hierarchic activities and different iterations of real projects. Concluding the paper, Section 2.5 discusses further work for a successful implementation of the iPeM.

2.2 Integrated Product Engineering Model

The integrated product engineering model (iPeM) was introduced as a modelling approach to depict product engineering processes, based on five hypotheses about product engineering (Albers, 2010). This paper threads the first, second and third hypotheses: *uniqueness of engineering processes*, the *triple system of product engineering* and the central activity *validation*.

2.2.1 Literature Review

An extensive derivation of the iPeM and the related state of the art can be found in Albers and Meboldt (2007) and Albers (2010). With the iPeM, engineering processes are described as a superordinate system consisting of an operation system that continuously generates objectives and transforms these into a system of objects.

This perception traces back to Hubka and Eder (1988) and Daenzer and Huber (2002). The systems engineering approach has been transferred to product engineering (Ropohl, 1979; Ehrlenspiel, 2003; Sim and Duffy, 2003).

There is no single approach commonly accepted as concurrently capturing all sorts of possible design situations while being specific enough to provide support on an operative work level today (Clarkson and Eckert, 2005). See the work of Browning and Ramaseh (2007), Wynn (2007), Meboldt (2008) or Albers *et al.* (2008) for a more detailed literature review on product engineering processes and process models.

2.2.2 Meta Model of Product Engineering

Figure 2.1 shows the meta model of the iPeM with its subsystems. The operation system is shown in the centre of the figure and contains the activities of product engineering (*macro* activities) and the (*micro*) activities of problem solving.

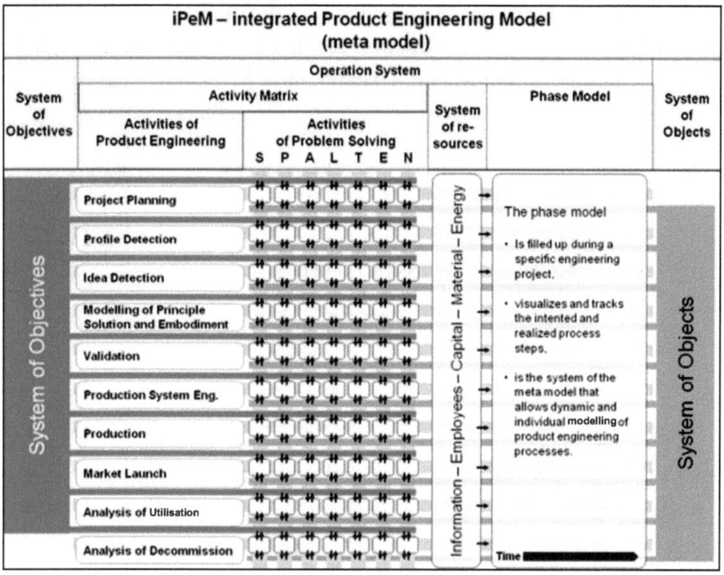

Figure 2.1. Meta model of product engineering (Albers, 2010)

Together they form the activity matrix where operational product engineering takes place. Using sets of generic activities as a basis for modelling processes allows a consistent and coherent description and facilitates a common understanding of a process in a group (Sim and Duffy, 2003). The picture shows the interface to the system of objectives on the left and the system of objects on the right.

To be able to consider organisational aspects of engineering processes as well, the system of resources and the phase model, where processes can be planned and tracked completes the operation system (for further explanation see Albers, 2010).

2.2.3 Abstraction Levels of the IPeM

Reasons for the fractal character on different levels of abstraction can be found in systems theory and the fundamentals of modelling. Models can be differentiated by their level of abstraction of formal structure as well as by their level of individuality (Rupprecht, 2002). Ropohl deduces the hierarchic dimension of the triple system based on the model theory by Stachowiak (Ropohl, 1979).

Concerning the formal structure, the *meta model* of the iPeM provides the basis to derive more specific *reference models*. These describe patterns of successful processes. They constitute the basis for formulating *implementation models* for specific projects. *Application models* monitor the actual course of a process for deriving information about improvement potentials for further reference models (Albers and Muschik, 2010a). The different abstraction levels of content-based individuality for an application of the iPeM in research are described as the *general* view, *domain-specific* level and *product-specific* level. For the application in industry, the second content-based level can be substituted by the *company-specific* level. Models for the different abstraction levels can be derived through two main procedures: *induction* or *deduction* (Albers and Muschik, 2010a) - see figure below.

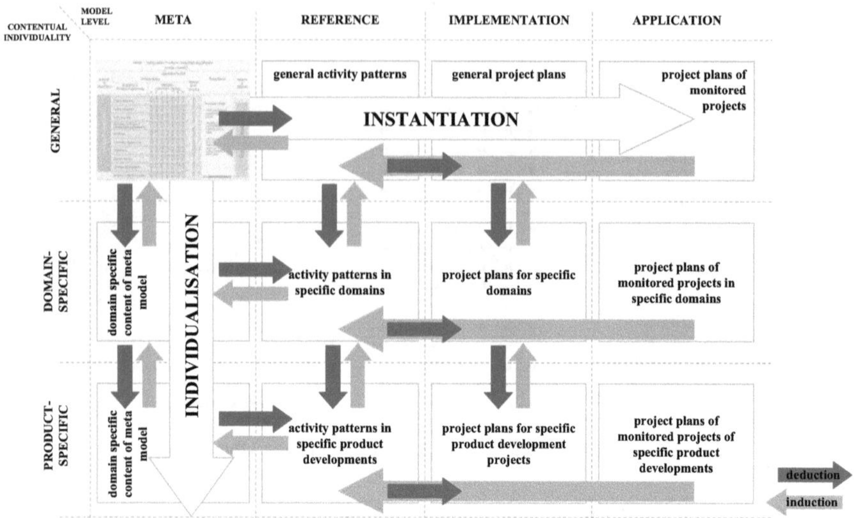

Figure 2.2. Abstraction levels of the iPeM (research application) (Albers and Muschik, 2010a)

2.3 System of Product Engineering

The following picture illustrates the interdependencies between the system of objectives, the operation system and the system of objects.

Initial objectives are derived from early information (*e.g.* about the market situation or competitor's activities) by the *operation system*. On this basis new objectives are generated continuously by forming and successively enhancing the *system of objectives*. The necessary activities are contained in the activity matrix to which elements of the system of resources are assigned. Through the activities of engineering processes, objects are produced, *e.g.* drawings, information documents, and plans *etc.,* which build up the *system of objects*. This influences the operation system in turn, leads to new objectives or can - in the case of a negative validation result - cause iterations in the engineering process.

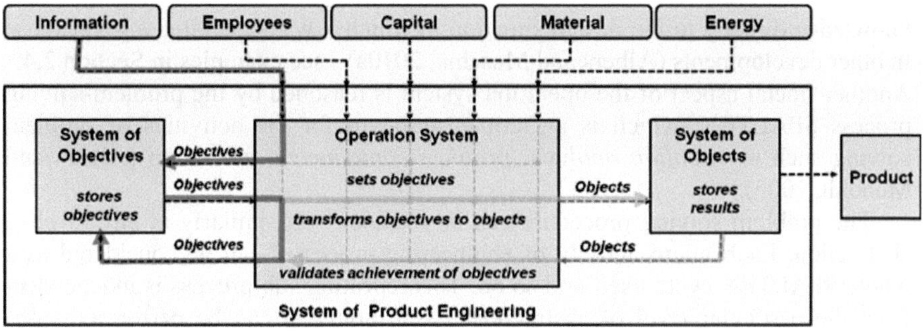

Figure 2.3. System of product engineering (see also Albers, 2010)

2.3.1 Operation System

The operation system incorporates all necessary means for the transformation of the system of objectives into the system of objects (Ropohl, 1979). Apart from the necessary resources such as labour, funding, equipment and supplies *etc.* the iPeM models the operation system by means of the activity matrix (see Section 2.2.2).

The activities can be explicated as systems that transform an input into an output according to the functional concept of a system (Albers and Muschik, 2010a). This output is again input to other activities leading to various interrelationships within the activity matrix. The input and output flows - so-called *deliverables* - lead to a hierarchic meshwork of different aggregation levels and differing granularity (Browning *et al.*, 2006).

2.3.1.1 Fractal Dimensions of Operation Systems

The activity matrix with its manifold interrelationships is of a fractal character. The iPeM's understanding of activities in the context of modelling product engineering processes is based on Hubka and Eder's hypothesis that designing is a rational cognitive activity, which is decomposable into smaller steps, *i.e.* into different levels of abstraction to match the relevant design problem. They propose a hierarchy of design activities, from design stages and according operations to basic operations, conducting a problem-solving cycle down to the lowest level of elementary activities and operations (Hubka and Eder, 1996). Every activity on one level comprises the activities on the following lower level and is itself part of all of its superior activities (Albers and Muschik, 2010a).

It is vital to understand that the set of activities in the meta model of the iPeM is dynamic, *i.e.* it develops according to the knowledge of product engineering processes. Therefore, it is never completed. An information transfer about occurring activities, their attributes and relations must be ensured to keep the model complying with changing constraints in continuously developing product engineering processes.

As the set of activities is adaptable to different abstraction levels, the iPeM can support modelling all kinds of design situations in order to handle the complexity of product engineering processes (see Section 2.1). It is possible to assign knowledge directly to the design situation in which it was generated *e.g.* for reuse in other developments (Albers and Muschik, 2010a) - see examples in Section 2.4. Another fractal aspect of the operation system is reasoned by the problem-solving process SPALTEN, which is a German acronym for the activities of problem solving such as *situation analysis*, *problem containment* and so on (Albers and Meboldt, 2005).

The problem solving procedure can be repeated self-similarly at any level of abstraction. Each micro activity of engineering processes can be considered as a whole SPALTEN cycle itself and so on. This repeating sub-process is independent from the particular level of abstraction. Even though it can be performed case-relatedly (which means that not every SPALTEN-step necessarily needs to be taken) the overall procedure is similar for every engineering activity in operational practice.

2.3.2 System of Objectives

A system of objectives contains all objectives, the boundary conditions and relations between them relevant to the development of a product. Equivalently to the operation system it is built of multiple subsystems. These cluster the elements of the system of objectives in so-called objective sections with respect to their content-based affiliation (*e.g.* objectives concerning technical, financial or social aspects) (Albers and Muschik, 2010b). Elements are information about boundary conditions relevant to an objective and objectives themselves (*e.g.* "weight has to be decreased by 20 kg"), expressed through tendency (*"to be decreased"*), parameter (*"weight"*) and value (*"20 kg"*).

Systems of objectives develop dynamically. During an engineering process information is allocated to respective objective sections and elaborated until a profound basis for the definition of an objective exists. A decision, *i.e.* setting an objective, is always a subjective procedure conducted by an agent. Consequently, at any point in time a system of objectives contains objectives in different development states (Albers and Muschik, 2010b).

2.3.2.1 Influence on the Uniqueness of Product Engineering Processes

The execution of a certain activity in the engineering process never proceeds equivalently in different projects. One reason for this is the evolved state of the assigned system of objectives that is regulating the sub-activities necessary for the transformation of objectives into objects. It can be differentiated from its previous states by different sub-activities that are necessary for the transformation, by the elements added, deleted or changed (*e.g.* tendency of objective or value).

A cause for this development of systems of objectives is the *change of boundary conditions* to objectives. These can be induced exogenously, *e.g.* through changes in society's trends. A rising consciousness of the environment might be relevant for a new product generation or changes of competitors' strategies might be important during a product's development. Changes can also be evoked endogenously, *e.g.* by changes in a company's financial situation. Another reason is the *subjectivity* of each *decision* necessary for the setting of an objective.

Different agents might derive different conclusions from boundary conditions for the definition of an objective. Also the *gain of insights* from further development projects *i.e.* from validating objects with according objectives during an engineering process causes changes in the systems of objectives.

Additionally changes to elements cause relations between them to develop. For example, the combination of two technologies in one product that has not been possible in a previous product engineering process might be technically feasible in the generation of a new product. Overall consistency of the elements might also be affected.

Decision situations for the definition of objectives, prioritisation of objectives in the engineering process as well as their value and consistency depend on the development of boundary conditions, performing agents and state of insight into a product. Therefore each system of objectives is unique.

2.3.2.2 Fractal Dimensions of Systems of Objectives

This uniqueness of systems of objectives aggravates its modelling and management for product engineering. The iPeM approach tries to overcome this difficulty in describing the fractal dimensions of the system of objectives on different abstraction levels. One product's system of objectives can be seen as a subsystem to the system of objectives of a company, which itself is subsystem to the system of objectives of society. The product's system of objectives can be subdivided into systems of objectives for its components and its parts *etc.* This is the content-based individualisation of the meta model for a system of objectives (see Figure 2.1).

Each of these models can be concretised with regard to the content of its elements (instantiation). Whereas the system of objectives on the meta level contains the main building blocks for objective sections and elements, a reference model specifies the relevant objective sections, *i.e.* elements, tendencies and relations for process patterns known from experience. Specific values for objectives are described on the implementation level.

These models on the different abstraction levels are derived deductively or inductively from project insights. Each model of the system of objectives triggers activities of the operation system which transforms the objectives into the system of objects by applying the problem solving steps in the product engineering activities - no matter on which abstraction level.

2.3.3 System of Objects

The result of product development processes are products. Nevertheless much more output is generated during the transformation of a company's system of objectives. There are three main output classes:

- (physical) objects that constitute the system of objects;
- information which is stored in the system of objectives;
- expert knowledge and know-how in the operation system.

2.3.3.1 Fractal Dimensions of Systems of Objects

As stated in Section 2.3.1, each activity can be considered as a system that transforms a system of objectives into a system of objects. Since the system of objectives can be modelled hierarchically in different levels of abstraction, also the resulting system of objects is a hierarchy with different abstraction levels and consists of individual objects that are correlating with each other through various relations. Especially design changes and iterations during engineering processes lead to individual cross-references between the system's elements and cause distinct systems of objects. Objects that are usually generated by activities later in the life cycle such as in the activity *production system engineering* need to be generated and essayed in early engineering stages such as *modelling of principle solution and embodiment*: the minimal radii of milling tools determine minimal radii of inner contours and have to be known before the actual life cycle phase is reached. Therefore, it can be necessary to jump from one macro activity of product engineering to another when additional information is needed and requires the analysis of certain objects *e.g.* in order to find out the minimal radius for a curve as exemplified above.

Another cause for iterations and design changes in particular are negative results of validation activities. On the one hand we find the macro activity *validation* in the iPeM. It represents engineering activities such as digital or physical prototyping, testing or final inspection. The object *product* is validated which means that it is verified according to the requirements that have been documented in the system of objectives and stored as information objects during the early engineering activities (Oerding, 2009). Objects are validated with respect to other objects.

On the other hand, objects are validated in reference to themselves or better: in reference to the information that is gained during their development's micro activities. During SPALTEN, activities such as *analysis of consequences* are performed. This step can be considered as a whole SPALTEN-process at a deeper level of abstraction starting with *situation analysis etc.* The object(s) created at this level influence the validation of superordinate objects.

2.4 Application in Wiki-based Implementations

The immense growth of (cross-linked) information that is produced during today's product engineering processes and the vast amount of knowledge in the hierarchic levels of abstraction that has to be handled by operation systems challenges product development. Useful tools are needed that help to manage this knowledge and to organise engineering processes. The iPeM as generic engineering process model can be considered as a universal structure for documenting and accessing this knowledge. In this section we present two approaches designed as first implementations of the iPeM in order to explore its applicability to real world processes. The exemplary implementations illustrate the advantages for structuring processes and point out weak points when applying the approach which are to be addressed by future research.

In these primary implementation cases the meta model was used to build implementation models. After completed projects, the resulting application models shall be used for the development of reference models for future engineering processes.

2.4.1 Implementation in an Industrial Environment

The implementation and validation of the expounded approach took place in cooperation with one of the leading solution providers for the print media industry using a SemanticMediaWiki with Halo-extension.

The implementation is characterised by four elements. Firstly, the documentation is designed referring to the iPeM approach with *separated but cross-linked systems of objectives and objects*. Secondly, these systems' *evolution can be tracked* during the engineering process by a continuous representation of the project's actual state. The third element is a functional-based subdivision of the necessary activities leading to a *hierarchy of levels of abstraction* as outlined in Section 2.3. Finally, *iterations are captured* by documenting the affected activities with the ambition of creating a fictive linear-chronological development process (Albers *et al.*, 2010c).

The application in the industrial context revealed several advantages and possible problems. The clear separation of objectives and objects helped to develop the correct thing in the correct way. The explicit linking structure allowed a comfortable and efficient change of perspectives between the objectives and the progress of the project respectively the latest version of the system of objects. The cross-linking of the hierarchic, functional structure helped to avoid interface problems between the systems and their corresponding sub-/super-systems. By the realisation of the knowledge base and the project documentation (in this case a product development process) within one (wiki-) system, a smooth and easy exchange of information *i.e.* knowledge was provided.

Problems were identified concerning the general acceptance *i.e.* comprehension of the iPeM as a mental model and its contained aspects such as system of objectives and objects and the activity matrix. It is sometimes not clear when and under which circumstances a system needs to get divided into several subsystems. It was questioned how much time should be spent on documenting and transferring the project's information into the knowledge base. An important conclusion was that the organisational culture of a company needs to accept and live for the approach of sharing knowledge (Albers *et al.*, 2010c).

2.4.2 Application in an Undergraduate-student's Project

The second exemplary application takes place in an undergraduate-student's project. In a team of four research assistants and 13 students of mechanical engineering an innovative automobile has to be designed. This engineering process started from scratch by identifying the market demands and will be accomplished with the roll-out of a prototype car.

Considering the aspect of knowledge management the challenge of this project is the huge amount of distributed information to be organised in one common document in order to capture the team's knowledge. Not only current work has to be supported but also future projects at the institute will benefit from the coherent knowledge base being generated during the engineering process.

In this application a semantic wiki system has also been chosen. The core idea of the representation is a twofold access to the knowledge base. In terms of the system of objects, the subprojects' information bases are arranged and represented in the order of their hierarchical level in the project. The documentation of the project's actual state can be described by the users directly in their subproject's workspace using the taxonomy of the iPeM. At the same time, the corresponding systems of objectives are connected via hyperlinks that enable users to switch to the description of their individual objectives as well as to the neighbouring constraints by a single mouse click (See Figure 2.4).

Before the application in the undergraduate-student's project an unstructured wiki system has been used as a platform for the students to share their contents. A survey revealed the need for an implementation of a meta model such as the iPeM that provides a structure in which designers are able to case-relatedly share knowledge and cooperate. The presetting of the formal structure of the iPeM *top down* in cooperation with the extension of the structure *bottom up* was due to project insights that the students found the deductive/inductive modelling approach from Section 2.2.3 to be helpful in operational use. Further results will be investigated in a second survey in the near future. The gained insights shall then be used as requirements for a revised implementation approach of the iPeM to be established in future work.

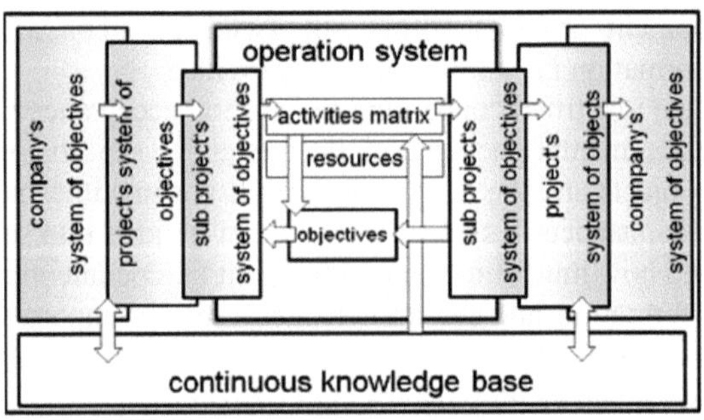

Figure 2.4. Layout of the wiki system in the student's project

2.5 Further Work and Conclusion

Today's product engineering processes are characterised by a huge complexity that amongst others can be explained by the multiple fractal character. One promising approach to handle this complexity is the model approach of the iPeM because it captures the different levels of abstraction of engineering processes. The application of the iPeM in wiki-based implementations has been suggested and exemplified. The two examples showed the general feasibility of the approach but also revealed weak points of current implementation approaches.

The pilot applications showed a basic practicability of the iPeM but also that further refinements of the approach are necessary. Changes in processes, organisational culture and parallel education about the use of the model are important to improve its benefit. However, the possibility to describe the gap between product life cycle activities and detailed engineering process steps through abstraction levels of activities has been shown. A revised implementation of the iPeM that satisfies the requirements that have been accumulated in the previous sections can lead to a concept that allows the approach to enter engineering processes in industry on a large scale. A new implementation of the approach that takes the findings of the exemplary applications into account is scheduled.

This paper develops the idea of using activities on different abstraction levels for modelling the transformation of systems of objectives into systems of objects. The main focus lies on the detection of multiple reasons for the fractal character of these levels. As self-similar structures could be verified, knowledge-management approaches that implement the fractal structures of the subsystems of the iPeM appear to be suited to handling the rising complexity of product engineering processes which has been affirmed in two exemplary applications. Further work will lead to an extensive and comprehensive implementation that will be used to validate the approach in sizeable industrial or educational projects with the overall aim to establish the applicability of the iPeM approach in the industry.

2.6 References

Albers A, Meboldt M (2005) SPALTEN problem solving methodology in the product development. In: Proceedings of the 15th International Conference on Engineering Design (ICED'05), Melbourne, Australia

Albers A, Meboldt M (2007) iPeMM. Integrated product development process management model, based on systems engineering and systematic problem solving. In: Proceedings of the 16th International Conference on Engineering Design (ICED'07), Paris, France

Albers A, Meboldt M, Deigendesch T (2008) Handling complexity - a methodological approach comprising process and knowledge management. In: Proceedings of the 7th International Symposium on Tools and Methods of Competitive Engineering (TMCE 2008), Kusadasi, Turkey

Albers A (2010) Five hypotheses and a meta model of engineering design processes. In: Proceedings of the 8th International Symposium on Tools and Methods of Competitive Engineering (TMCE 2010), Ancona, Italy, in press

Albers A, Muschik S (2010a) The role and application of activities in the integrated product engineering model (iPeM). In: Proceedings of the 11th International Design Conference (DESIGN 2010), Dubrovnik, Croatia, in press

Albers A, Muschik S (2010b) Development of systems of objectives in early activities of product development processes. In: Proceedings of the 8th International Symposium on Tools and Methods of Competitive Engineering (TMCE 2010), Ancona, Italy, in press

Albers A, Ebel B, Sauter C (2010c) Combining process model and semantic wiki. In: Proceedings of the 11th International Design Conference (DESIGN 2010), Dubrovnik, Croatia, in press
Browning TR, Fricke E, Negele H (2006) Key concepts in modeling product development processes, Systems Engineering, 9(2): 104-128
Browning TR, Ramaseh RV (2007) A survey of activity network-based process models for managing product development projects. Production and Operations Management, 16(2): 217-240
Clarkson PJ, Eckert C (2005) Design process improvement - a review of current practice. Springer Verlag, London, UK
Daenzer WF, Huber F (2002) Systems Engineering. Verlag Industrielle Organisation, Zürich, Austria
Ehrlenspiel K (2003) Integrierte produktentwicklung: Denkabläufe, methodeneinsatz, zusammenarbeit. Hanser, München, Germany
Hubka V, Eder E (1988) Theory of technical systems. Springer, Berlin, Germany
Hubka V, Eder WE (1996) Design science: Introduction to needs, scope and organization of engineering design knowledge. Springer-Verlag, Berlin, Germany
Meboldt M (2008) Mental and formal modeling, a contribution to the integrated product development model (iPeM). Institute of product development Karlsruhe (IPEK), Karlsruhe, Germany
Oerding J (2009) Ein Beitrag zum modellverständnis in der produktentstehung. Strukturieren von Zielsystemen mit Hilfe von C&CM. Institute of product development Karlsruhe (IPEK), Karlsruhe, Germany
Ropohl G (1979) Eine systemtheorie der technik: zur Grundlegung der Allgemeinen technologie. Hanser, München/Wien, Germany
Rupprecht C (2002) Ein Konzept zur projektspezifischen individualisierung von prozessmodellen. Dissertation, University of Karlsruhe, Karlsruhe, Germany
Sandmeyer P (2009) Christentum - die lehre der liebe. Stern Extra 5: 26-43
Sim SK, Duffy AHB (2003) Towards an ontology of generic engineering design activities. Research in Engineering Design, 14: 200-223
Wynn DC (2007) Model-based approaches to support process improvement in complex product development. PhD thesis, Cambridge Engineering Design Centre, University of Cambridge, Cambridge, UK

Chapter 3

Approaches for Process Attendant Property Validation of Products

K. Paetzold, J. Reitmeier

3.1 Introduction

Continuous validation of product functionality must be ensured throughout the development process. In terms of virtual product development, this should be done without physical prototypes, or at least their use should be reduced to a minimum. Simulation methodology is an appropriate device to gain knowledge about real systems through models. A multitude of highly efficient simulation tools for different disciplines is available today. It is still an open question which simulations can be executed at what point in time to really support product development processes in terms of concurrent engineering to reduce the development risk by purposeful and early validation of product functionality.

This paper aims to show aspects to be considered when integrating simulations into the development process in order to prepare efficient simulation planning.

3.2 FORFLOW-process Model as a Basis to Describe Development Processes

If continuous validation of product properties is to be ensured throughout the development process, project planning (milestones, costs, resources) needs to include simulation planning too. The latter is integrated into project planning because simulations are usually carried out by specialists and require special tools, both of which can be considered limited resources. Furthermore, data quality must be pre-defined with respect to both completeness and certainty if simulations are to be executed efficiently. It is closely connected to the progress of the process.

Therefore, considerations concerning simulation planning require a detailed process model to effectively integrate simulation methods into the development process. For this purpose, the process model developed by FORFLOW (Bavarian research alliance consisting of six institutes of mechanical engineering and applied informatics of four Bavarian universities) may serve as a basis. This research alliance was the result of an initiative by several industrial partners and makes it possible to continuously validate research results by practice-related application scenarios. Constant positive feedback from industry partners has led to deeper evaluations, which are still ongoing (Krehmer *et al.*, 2009). The process model that has been developed is a detailed and variable, but nevertheless globally oriented approach to meet current challenges of product development such as the integration of mechatronic aspects or situation-specific process planning (Krehmer *et al.*, 2009). The core objective was to provide optimum process and workflow support to the developer in order to make procedures in the product development process more effective and efficient.

3.2.1 Basic Design of the FORFLOW-process Model

The various requirements to be met by modern technical systems not only lead to more and more complex products but also to an increasing complexity of development processes. Such process complexity is further intensified by the need to cooperate in multidisciplinary teams, which is a result of the trend towards mechatronic products and corporate-wide value-added chains and development processes. It is therefore important to have a process model whose basic strategic framework is fixed but which also allows for enough operational flexibility to respond to enterprise-specific or situational given factors. The FORFLOW process model (FFPM) provides such a process framework (Figure 3.1.).

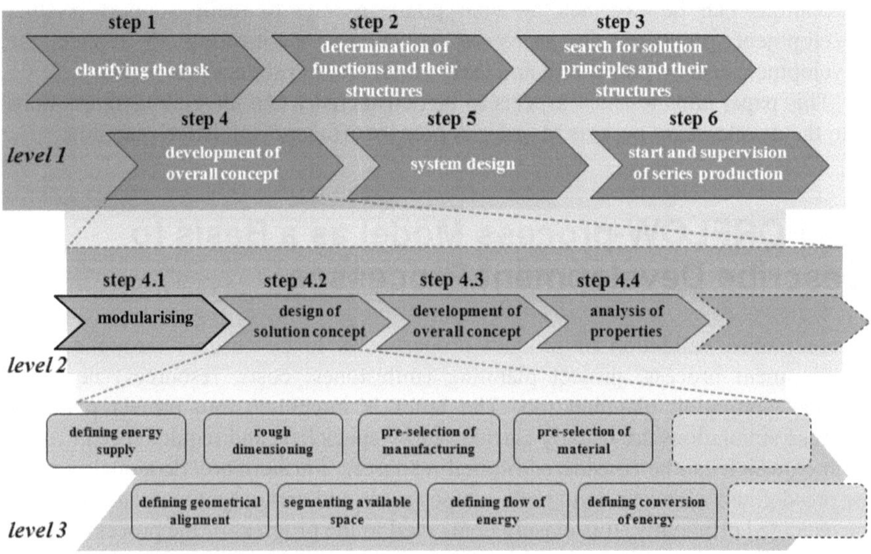

Figure 3.1. Three-level-design of the FORFLOW-process model (Krehmer *et al.*, 2009)

On the highest level, the product development process is roughly described by six main steps, *i.e.* conventional procedure models like VDI 2221 are extended by the step "start and supervision of series production" to reflect the industrial standard of series production support once production has started. This ensures that development-relevant information from production can flow back into the development process. Based on the structure of the company and its trade or branch of industry, the development process will be detailed further on the second level, which allows a classification of process steps into "optional", "mandatory", and "required". This facilitates adjustment to enterprise-specific conditions and to special given situations of development. On the last level, some process steps are split up further so that the development process finally comprises more than 90 individual steps, an obviously much more detailed representation than in the case of conventional process models (Paetzold, 2009).

The process steps on the first and second levels have to be executed in a given order, whereas on the third level, there is no restriction concerning the chronological process order. Thus, the process model on the one hand provides a realistic construct, but on the other hand offers enough freedom for process design, which will finally be determined by concrete development situations.

3.2.2 Particular Benefits of the FORFLOW-process Model

Flexibility is one major benefit of the FFPM; process flows are not fixed entirely, but can be adapted to some degree. This is indispensable if the process flow needs to be adapted as part of a concrete development process, or if insights during the development process suggest the requirement to adapt the planned process steps. In both cases, adaptation takes place on the second level, with process steps being categorised into "mandatory" and "recommendable" for the purpose of product validation. As already mentioned before, the steps of the third level need not follow a fixed chronological order, which provides the designer with enough freedom to adapt process steps to the requirements of his daily work. This process planning can be supported by the Dependency Structure Matrix (DSM), which shows how individual steps influence each other while taking into account existing general conditions and recommends an order of process steps (Roelofson *et al.*, 2008). As a rule, process steps with close reference to each other are executed simultaneously, others will be carried out sequentially. This is especially important in mechatronic product development where not individual steps, but the interaction of development processes of different disciplines, in connection with iterations, leads to an optimum solution.

In addition to product complexity, reasons for iterations may be wrong decisions, faulty communication or changing market and / or consumer requirements. Furthermore, every iteration contains decisions or partial results that have to be documented. For this purpose, the FFPM provides multiple process interfaces where the process flow can be left. At the end of a relatively long process step, product-related results as well as the engineers' lessons learned about the process flow are documented and made available for subsequent projects.

A highly flexible development process like this makes it necessary to integrate dynamic product models, which must be connected to the process. Conventional process models either use the product structure or do not allow any flexible modifications. In order to provide desired product information anytime, a parameter-based description method for product models was designed (Lauer *et al.*, 2008). This approach makes it possible to automatically assign product models to individual process steps according to their relevance. Thus, a flexible connection is guaranteed, which in turn ensures an optimum flow of data.

Product models are the result of the computer-aided tools (CAx tools) integrated into the development process. Whereas the FORFLOW project primarily focused on questions concerning data compatibility and modelling guidelines, the aim of the following exposition is to integrate simulations into the process according to the expected property and the special product functionality. Tools available today on the one hand are characterised by high performance, but on the other hand require a high degree of sensitivity with respect to the necessary input data quality and the informational value of output data. Established companies are known to have an abundance of experience in dealing with questions that concern the correct application of certain CAx methods. This knowledge is documented in guidelines, which, among other things, may also contain legal provisions. This ensures that information can be provided as required and results can be evaluated in an optimum way any time.

Another aspect looked at as part of the FORFLOW project was the selection and application of available CAx tools depending on organisational framework conditions such as "availability at suppliers" or "in-house competence and capacity". Related guidelines and recommendations for actions were integrated into the process model on the basis of connections between process steps and product models and / or the tools these product models are based on. In addition, a classification of the CAx tools facilitates integration into the development process.

3.3 Holistic Approach to Simulation Planning

3.3.1 Phase Assignment in Terms of Execution of Simulations

With the help of the fine-grained process model explained above, the data needed for simulation can initially be captured and structured very well due to a detailed description of the input and output values of the process steps. However, there is no information as to which kinds of simulations are required and can be executed considering the properties to be validated and the data available, and there is no indication at which time during the process the individual simulations should ideally be triggered. Finally, simulation planning also means that for the third level of process planning described in Section 3.2, information can be derived as to which simulations have to be evaluated how in terms of the classification of process steps ("required", "mandatory", and "optional" steps). Considerations about simulation planning are based on the assumption that development processes have to be understood as a permanent interplay between synthesis and analysis processes. If product characteristics are determined during synthesis, an analysis needs to be conducted in order to verify whether the properties resulting from these specified characteristics really conform to the requirements. Such characteristics describe the structure, shape or material consistency of a product and can be directly influenced by the designer (material, dimensions, *etc.*). Properties describe the product's behaviour (weight, reliability, *etc.*) and cannot be influenced directly. Thus, simulations of any kind have to be regarded as steps of analysis whose execution depends on the availability of characteristics. Accordingly, the development process outlined can be subdivided into three phases which are briefly described below.

3.3.1.1 Determination of Target Properties of a Technical System

In the early stages of the development process, the main focus will be on the definition and detailing of the properties the future product should exhibit. In a first step, all requirements (customer requirements, design for X (DFX) criteria, boundary conditions) must be captured and described so that they can be used as comparison criteria for the validation process. This means that during the individual work steps, properties are laid down which specify and detail the developing product. The granularity of this detailing must make it possible to validate the properties in the course of the development process by means of simulation. According to the top-down approach in the development, a hierarchical structure of these properties is to be expected. This approach starts with customer requirements, which are specified, detailed and described in the requirements specification generated for that purpose. Already at this point, it becomes obvious that a purely hierarchical description is insufficient because the required properties interact. However, this aspect is not considered more closely in the early stages of development in order to guarantee the necessary neutrality of results (Pahl *et al.*, 2007).

3.3.1.2 Domain-specific Determination of Characteristics of Single Components

In the subsequent stages of domain-specific design, characteristics are defined which contribute to the fulfilment of the properties. The definition of characteristics begins with specifying the concept. In general, the chosen solution principle determines the distribution of tasks among the individual domains of the engineering disciplines. Usually, the sum of characteristics leads to new and/or higher-ranking characteristics, *i.e.* a bottom-up strategy is usually pursued in their development. The synthesis steps described for the determination of characteristics always include an analysis, too, *i.e.* a verification whether the characteristics fulfil the desired properties. Two aspects are typical of this stage:

- usually, property verification is executed locally, *i.e.* within the departments. Relations and interfaces between different domains are considered in terms of boundary conditions, if at all;
- the definition of characteristics also affects properties and causes properties to occur which were not in the focus of the synthesis steps but which can

be identified by effective methods of analysis and therefore must be considered.

3.3.1.3 Description of Properties of the Overall System

In the third stage, consolidation and optimisation of properties take place as the partial solutions and/or partial systems that have developed in the individual domains have to be integrated in an overall system. Now, the actual properties of the developing technical system are focused on. This is associated with an aggregation of properties, which is important for property validation (a bottom-up procedure). Individual properties such as the strength of a component or the rigidity of a casing are used to derive higher-order properties such as noise emission, serviceability or handling. Due to the fact that properties may influence each other, the structure of the actual properties should not be expected to be purely hierarchical but will rather resemble a network.

The properties determined during the early stages serve as comparison criteria for the actual properties. In this context, the characteristics connected with the actual properties may serve to manipulate the latter to a certain degree in order to adapt the construct of reference properties by modification of the actual properties as best as possible.

3.3.2 Classification of Simulation Methods

To optimise functional validation by simulation, it finally needs to be clarified whether every simulation method may be used at any time during the development process, which prerequisites have to be linked to its usage, which results and the result quality can be expected. For this purpose, the characterisation of product data (properties/characteristics) described in the above chapter, as well as the aspects “uncertainty” and “incompleteness” of product data must be considered during the development process.

Simulation methods may be subdivided into two main categories: parameter-oriented approaches and field-theoretical approaches (Bossel, 1994) (Panreck, 2002). Parameter-oriented simulation approaches imitate the behaviour of the system to be simulated. The system model, *i.e.* the actual structure (*e.g.* geometry in mechanics) itself is of relatively minor importance, while it is essential that the behaviour of the model corresponds to the original. In this context, the model of the system is based on physical principles which are chosen as sub-functions. The sub-functions will be connected by using material, energy and information flows to fulfil the overall function. Due to their focus on system behaviour, these approaches can be described as black-box methods. Usually, field-theoretical approaches pursue the goal to represent the system structure in simulations. Therefore, the structure of the original must have been identified and understood. Accordingly, these approaches are also referred to as glass-box methods.

When taking a closer look, it becomes clear that the data required for both simulation methods must meet very different demands. This will be explained by looking at the stages of the development process from Section 3.3.1.

3.3.2.1 Simulation Approaches within Concept Planning

During the process steps "concept design" as well as "development and completion of the properties of the product to be developed", only little information will be available in general. This information will frequently have to be regarded as uncertain, since there is no concretisation. Information about the structure is not available; although it is being prepared in the concept development stage, its concretisation is a responsibility of the specialists of the domains in the subsequent process steps. Therefore, glass-box approaches are of no significance at this development stage. It is important that the concept of a product and the chosen general solutions are able to represent the expected system behaviour. For this purpose, black-box methods (*e.g.* Matlab-Simulink or Dymola) provide excellent simulation methods.

Input and/or output values are the result of customer requirements, DFX criteria or business objectives. On the one hand, objectives for component development in the specialist domains can be deduced from coefficients and abstracted descriptions, on the other hand, this approach facilitates decisions concerning optimised solutions as the modelling occurs in a domain-neutral form and guarantees a solution-neutral result. However, overall system behaviour is simulated with limited granularity. The latter depends on the granularity applied in describing the reference properties. Such considerations of the overall system are of major advantage since already at this stage, interfaces and solution principles for the associated components become obvious and can be communicated.

In spite of the mentioned advantages, it must always be considered that the data available for simulations at this stage are incomplete and uncertain. They primarily serve to concretise the developmental task. However, as far as the representation of the reference properties is concerned, they provide acceptable comparison criteria for the following process steps.

3.3.2.2 Simulation within the Scope of Domain-specific Concretisation

During the stage of the detailing and elaboration of the solution principles and components in the domain-specific departments, the focus is on the definition of characteristics and their detailing, with the required properties serving as a basis. As a result, information about the structure is generated, and it soon becomes evident that structure descriptions vary considerably among the domains. Whereas a mechanical engineer considers "structure" to be geometry, a computer scientist may regard source code as structure, and for an electrical engineer, this concept may denote circuit diagrams and electronic circuitry. Simulation methods must be able to represent these specific structures in order to establish property validation. Therefore, glass-box methods, such as finite element methods (FEM) are primarily applied at this stage. These simulation methods make it possible to identify the resulting properties for domain-specific characteristics and groups of characteristics. Usually, a comparison with properties is only partially possible, depending on the granularity of the description. In general, domain-specific guidelines or engineer standards are used as comparison criteria in addition to consumer requirements.

The disadvantage of these simulation approaches is that connections to other properties can be considered for comparisons only to a limited extent. This is caused by heterogeneous structure descriptions, *i.e.* in many cases, different terms are used for the same facts. If connections between properties are known from the description of the properties, they are integrated into the simulations as boundary conditions. Frequently, interactions between properties arise from simultaneous development in other domains. Assumptions are made which differ from the characteristics actually selected. This uncertainty in the connection of partial solutions will certainly affect the quality of results of the actual properties determined. In addition, the bottom-up approach described only allows to reliably determine actual properties on the lower levels of the network, whereas conclusions about the consequences for the overall system are possible only to some extent.

3.3.2.3 Simulation for Considerations of the Overall System

One of the major challenges in product development is the combination of the individual components to form an overall system. For simulation purposes, the structures (described by appropriate specifications of characteristics) which come from different domains - and thus vary - must now be combined to form an overall system whose system properties correspond to the defined properties.

A consideration of the overall system can be realised both by glass-box and black-box approaches. The use of simulation methods representing the complete structure description soon leads to problems, as structure definitions are very heterogeneous among the domains. This makes it necessary to couple the models or the simulation tools themselves, which is a very time-consuming and complex procedure. The error rate is high and, accordingly, the overall statement concerning the system function will be doubtful. Furthermore, such simulations not only require an enormous computer processing effort, but also a very high degree of know-how to reasonably interpret the simulation results.

It seems more useful to fall back on black-box approaches, which only require a description of the overall system's behaviour. Possible methods that may be mentioned here are model-in-the-loop (MIL), software-in-the-loop (SIL) and hardware-in-the-loop (HIL) simulations. Their advantage is that behaviour definitions are comparable among all domains. However, for this purpose, the description of characteristics for the components from the domains must be reduced to only include the most typical properties and characteristics, which can then be used as coefficients. A very decisive advantage of this method is the fact that the actual properties of the system can easily be compared with the initially defined properties. In addition, it becomes more transparent how partial solutions affect the behaviour of the overall system. Manipulation parameters are easier to recognise and thus, optimisation of the overall system is facilitated.

3.3.3 Aspects of Simulation Planning within Development Processes

3.3.3.1 Simulation Placement within the FORFLOW-development Process

The focus of the FORFLOW research project was, among other things, to establish the relevance of the integration of components into an overall system in order to consider interdependencies between components as early as possible and thus to also ensure product functionality. Although the FFPM is displayed as a sequential procedure, the aspects mentioned make it easy to convert it into the V model which is common for mechatronic systems in accordance with Directive VDI 2206. This view is essential to integrate simulation planning efficiently. Figure 3.2 shows the allocation of the six main steps of the FFPM as well as a phase allocation of previous simulation aspects with respect to this V model.

Starting from a concrete development order, the task, which depends on product requirements, must be clarified first. This starting point of a project is equal to step one of the FORFLOW process, "clarifying the task". Already here, initial important information concerning simulation planning must be documented. This may be the observance of legal regulations (*e.g.* crash tests in vehicle construction) in terms of "mandatory activities" and/or an initial definition of simulation validations based on lessons learned from similar development projects. Such provisions not only determine which simulations are required, but they also indicate target properties and comparison criteria. The defined product requirements and properties must be further described and detailed in a top-down approach. This procedure of detailing the system design with increasing granularity corresponds to FORFLOW step two "determination of functions and their structures" as well as to step three "search for solution principles and their structures". These process steps reveal analysis steps required for the property validation of partial systems and of the overall system.

Step four of the FORFLOW process starts with the domain-specific detailing and elaboration of solution principles and components. Here, the focus is on the definition of characteristics to obtain the reference properties identified before. This definition of characteristics is associated with a first definition of the domain-specific simulation validations. In general, specific standards and guidelines for the detailing process are known by the specialist disciplines. This implies further required and mandatory steps for property validation while also providing comparison criteria. Thus, the resulting domain-specific splitting of properties during the concretisation phase helps to keep an overview of processes. However, there is a risk that resulting interfaces to other domains or components are not communicated to a sufficient degree.

At step five of the FORFLOW process, the necessary interplay of components from different domains dealing with mechatronic products necessitates further simulations. Their primary aim is to capture undesired interactions that may have negative effects on product properties (*e.g.* electromagnetic compatibility) in order to make it possible to eliminate them and to ensure overall functionality. Before this point, initial comprehensive simulation planning cannot be completed. Even today, there are no really satisfactory methodical approaches to simulate overall systems, because models become very large and complex, no matter what simulation philosophy is adopted. An enormous computer capacity is required and system engineers are needed for interpretation who must be able to comprehend the products' complexity. Although MIL, SIL and HIL simulations provide a good basis for such purposes, they cannot be regarded as sufficient methods.

The last step "start and supervision of series production" examines an already marketable product. For example the possibility exists to integrate simulations into a digital factory and besides product optimisation with regard to manufacturing and quality, simulations and optimisations concerning the manufacturing process such as more efficient assembling, *etc.* can be successfully integrated.

Knowledge gained from initial analyses within the domain-specific design may necessitate modifications in simulation planning. Thus, it may be necessary to modify characteristics as well as the solution principles in order to obtain the desired product functionalities. This in turn will lead to iterations. Of course, such iterations or the decision to leave out or add property validations may also occur in the subsequent validations of partial systems and/or the overall system, since the simulation steps and simulation efforts required can only be roughly planned before. Their real procedure or composition in fact can only be formulated explicitly when looking at the latest results of analysis and the progress in development. It must be pointed out that a purely virtual property validation is not sufficient, and that especially on the overall system level, product functionalities have to be validated by physical prototypes. On the one hand, this is an inherent requirement of the computational procedures themselves, and on the other hand, it is often demanded by the legislator or the customer.

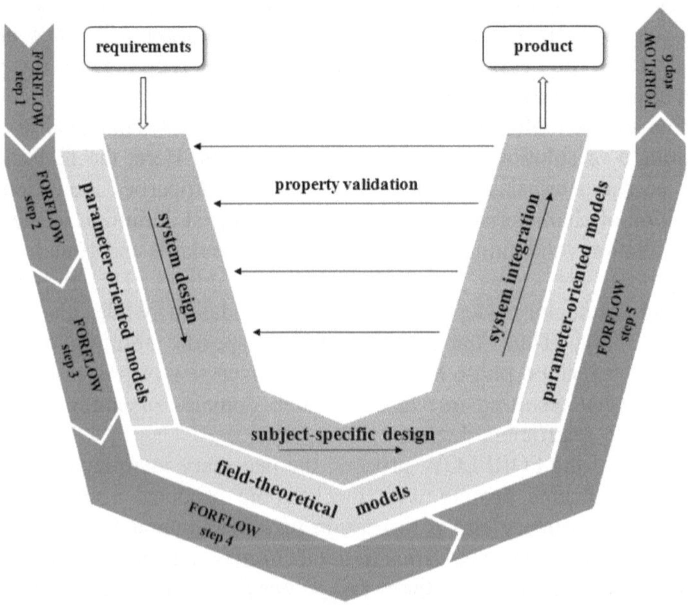

Figure 3.2. Simulation placement within the FORFLOW-development process

3.3.3.2 Resulting Necessities for Simulation Planning

The quality of simulation results and thus the obviousness of a simulation very much depends on the data available. Therefore, efficient simulation planning always requires an efficient provision and preparation of data, too. As process observation has shown, a clear differentiation must be made between properties and characteristics (Weber, 2005) in this context. Characteristics will always be the input for any simulation, whereas properties will be the simulation output. A clear categorisation of product data into properties and characteristics is neither done today nor supported by corresponding tools.

If simulations are to be employed reasonably, comparison criteria must be defined in order to verify simulation results. This, too, is a problem which has not yet been solved completely: Although customer requirements, DFX criteria or business objectives are available as starting points, these are often rather obscure and vague concepts such as “comfort” or “safety”. Translating them into concretely measurable or rateable parameters is difficult as these soft factors may be interpreted very differently depending on the corporate philosophy or the developers’ state of knowledge.

Data certainty and data availability are a major problem in the execution of simulations. Especially in the early stages of concretisation, a lot of assumptions have to be made because data are simply not available. Working with assumptions is not unusual and certainly legitimate, since it is generally based on the wide experience of developers. However, it will be absolutely necessary to indicate which input data the simulation results are based on (three broad categories seem sensible) in order to prevent systematic errors and avoid implying a precision that does not exist.

It was also mentioned in Section 3.3.3.1 that iterations are inevitable in the execution of simulations. This aspect must be expanded: detailed simulation planning in the phase before the development is neither possible nor wise. In fact, mechanisms are required which trigger simulations depending on the input data available. A simulation is meaningful only as soon as the necessary data of adequate certainty are available. Therefore, only one thing is specified on the basis of the required properties: the types of necessary simulations. The actual point in time when they will be carried out must depend on data release.

3.5 Summary

Simulation planning is necessary to ensure product functionality throughout the development process. The FFPM provides a framework which is detailed enough to assign questions of simulation planning to the development process more efficiently. In order to make it possible to develop simulation planning approaches, further research activities are being conducted which focus on the optimisation of data flows. From the authors’ point of view, this is essential in order to significantly reduce development cycles by means of the simultaneous validation of product functionalities in terms of concurrent engineering.

3.6 References

Bossel H (1994) Modellbildung und simulation - konzepte, verfahren, und modelle. Vieweg Verlag, Braunschweig, Germany

Krehmer H, Eckstein R, Lauer W, Roelofsen J, Stöber C, Troll A *et al.* (2009) Coping with multidisciplinary product development - a process model approach. In: Proceedings of the 16th International Conference of Engineering Design (ICED'09), Stanford, CA, US

Lauer W, Ponn J, Lindemann U (2007) Purposeful integration of product models into the product developement process. In: Proceedings of the 15th International Conference of Engineering Design (ICED'07), Paris, France

Paetzold K (2009) FORFLOW: Ein holistischer Ansatz zur Prozessunterstützung. In: Flexible prozessunterstützung in der produktentwicklung: Prozesse - daten - navigation. Shaker Verlag, Aachen, Germany, pp 25-42

Pahl G, Beitz W, Feldhusen J, Grote KH (2007) Konstruktionslehre. Springer Verlag, Berlin, Germany

Panreck K (2002) Rechnergestützte modellbildung physikalisch-technischer systeme. VDI-Fortschrittsberichte Reihe 8, Nr. 945. VDI Verlag, Düsseldorf, Germany

Roelofson J, Krehmer H, Lindemann U, Meerkamm H (2008) Using the Design-Structure-Matrix for the avoidance of unnecessary iterations. In: Proceedings of the 10th International DSM Conference (DSM'08), Stockholm, Sweden, pp 209-218

Weber C (2005) CPM/PDD - an extended theoretical approach to modeling products and product development processes. In: Proceedings of the 2nd German-Israeli Symposium on advances in methods and systems for development of product and processes, Berlin, Germany, pp 159-179

Part II

Managing Complex Engineering Processes

Chapter 4

An Approach Towards Planning Development Processes According to the Design Situation

J.M.K. Roelofsen and U. Lindemann

4.1 Introduction

The growing complexity of today's products and the need for a short time to market lead to an increasing complexity of product development processes. Thus, coping with this growing complexity of development processes becomes more and more important for companies in order to be able to face the international competition. Besides the market requirement for shorter product cycles, the increasing integration of electronics and software into formerly mechanic products adds to the growing complexity of design processes.

In order to meet these demands it is inevitable to deal with the process complexity and to be able to react flexibly to changes in the boundary conditions (*i.e.* the design situation) of product design.

This contribution introduces an approach towards the planning of product development processes according to the design situation that helps to enable companies in reacting to changes on the market, in customer needs and in other important constraints of product design. The different partial approaches, their integration as well as implementation and evaluation are discussed.

4.2 Motivation and Objectives

In order to cope with the growing complexity of product development processes, it is necessary to develop new methods for process management. This was confirmed by industrial participants of a workshop dealing with challenges in development process management. Existing approaches that support the management of well known business processes are often not applicable to development processes due to their specific characteristics. One difference between product development and other business processes is that the result of the process cannot be completely defined in advance (they are rather evolutional). Development processes are characterised by a high number of iterations (Vajna 2005; Wynn *et al.*, 2007) due to changes in product requirements or the fact that a requirement is not met by the solution developed first or due to changes in the boundary conditions of a development project, to name a few. The approach towards process planning introduced in this contribution supports both the reducing of iterations in design processes as well as the reaction to changes in the boundary conditions of a design project.

A more flexible planning of design processes leads to shorter development time and better product quality. The approach supports two different levels of process planning for development projects. These are the project and the operational level of process planning. It facilitates the instantiation of a generic process model for an oncoming development task and is thus applied at the interface of process and project management. Other project management tasks such as, for example, the assignment of resources to single tasks (Kerzner 2006), do not belong to the scope of this approach. Rather the main focus of the approach is to take into account the actual design situation (Maffin 1998) and to enable a flexible reaction to changes in this situation during a product development project. Therefore, the overall approach consists of a generic process model of the product development process, a description of the design situation and a procedural model for planning the process sequence of a development project. These are integrated in order to achieve an even better support for the product developing engineer.

The approach introduced in this contribution was developed in the research alliance FORFLOW, that consisted of six institutes from mechanical engineering and computer science and twenty affiliated companies. The goal of this alliance was to develop and implement new methods to support processes and workflows in planning and managing product development. This contribution is part of the research work carried out in close collaboration with the industrial partners.

4.3 Planning Development Processes According to the Design Situation

In order to describe the approach towards situation specific planning of product development processes it is necessary to introduce the elements the approach is composed of. This includes the process model applied in this approach, the different levels of process planning and respective roles as well as the description of the design situation and the basic procedural model. After introducing these basic ideas, the approach will be depicted and its evaluation discussed.

The goal this contribution aims at is supporting project planning as well as project executing engineers in planning their development procedures and thus reducing development time. Moreover, better process transparency leads to a better understanding of the developed system and support communication during the design process.

The approach is based on an analysis of development processes of the industrial partners and a collection of requirements for design process planning support derived from literature and discussions with the partner companies. The single parts of the approach and how they work together will be illustrated in the following.

4.3.1 The Process Model

The process model this approach is based on is the FORFLOW Process Model (Krehmer *et al.*, 2009). It is a generic model that consists of about 90 different process steps defined in three levels of detail (see Figure 4.1). A tailoring of this model is done for each new development project or within a project for different subsystems. The order of process steps in the first and second level of detail is predefined; the order of steps in the third level of detail has to be defined according to the development task at hand.

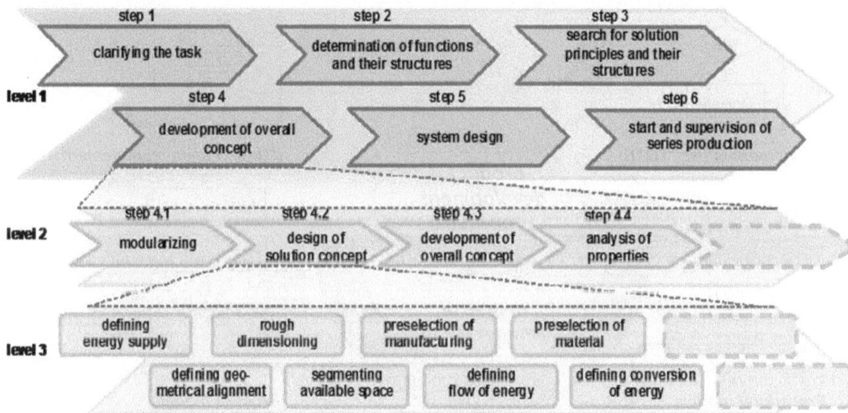

Figure 4.1. The three levels of detail of the FORFLOW Process Model

The FORFLOW Process Model was developed to meet important challenges in product development, amongst others the integration of mechatronic aspects, the integration of design iterations and multi level safeguarding.

This contribution focuses on situation specific process planning. A more detailed description of how the other challenges are met is given in (Krehmer *et al.*, 2009). The practical applicability of the FORFLOW Process Model was proven by the fact that tailored versions of the model can display the development processes of different companies affiliated with the research alliance FORFLOW. The FORFLOW Process Model was built to avoid the deficits of *e.g.* the process model of the VDI 2221 (VDI 1993), that cannot be adapted dynamically (Pesic and van der Aalst 2006).

4.3.2 Levels and Roles of Process Planning

In order to achieve the goals in product development concerning time, cost and quality it is necessary to look at the right level of process decomposition for the right task. According to (Lindemann 2007) there are four levels of decomposition and their respective process models (see Figure 4.2):

- strategic process level (with only generic processes and roadmaps);
- project level (with rough stages but a clear vision of outcome);
- operational level (with interrelated activities but vague certainty of final outcome);
- action level (with elementary processes).

The FORFLOW Process Model supports planning design processes on the project as well as the operational level (Roelofsen *et al.*, 2007a), as these are the levels of decomposition represented in the process model.

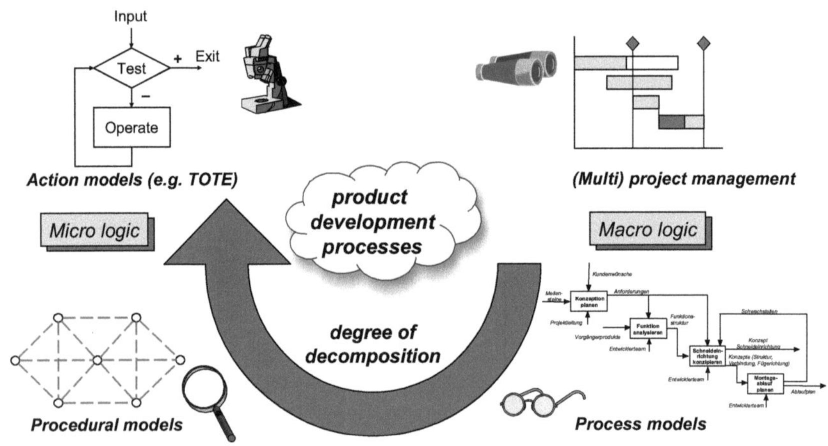

Figure 4.2. Design process models on different levels of decomposition adapted from (Lindemann 2007)

The approach presented here deals with two different roles in process planning according to the two levels of process decomposition that are supported. These roles are not directly connected to persons nor to a project organisation, in which more levels of hierarchy and a larger number of different roles can occur (Kerzner 2006). The differentiation between project and operational planning is sufficient to perform the kind of process planning presented here (see Roelofsen *et al.*, 2007a). This splitting of process planning into two roles makes the approach applicable for many different situations. It does not matter whether there is one project manager, who builds a project plan at the beginning of a project, or whether there is a group consisting of people from different domains that create the project plan in collaboration. The same applies on the operational level. Here, a working group can plan their tasks in order to develop *e.g.* a sub-assembly, or a single engineer plans and adapts the development of the product part he is responsible for. Finally it is possible that both roles are fulfilled by the same person, first planning the overall task and afterwards planning single steps in more detail.

The support on project level is implemented in the first two levels of decomposition of the FORFLOW Process Model. The project level support facilitates project planning for the respective role. The process model differentiates between three classes of process steps, according to their necessity in design processes. The necessary steps have to be chosen in project planning. First, there are steps that have to be carried out so that a producible product is the result of the process, *e.g.* the detailed design of product parts. These are the "MUST"-steps. Second, there are process steps that are recommended but not absolutely essential as for example working out a functional model of the product. These are the "SHOULD"-steps. And third there are process steps that can be carried out but are of less interest in some cases, *e.g.* the abstraction of the development task, which are the "CAN"-steps. The approach provides support to select and arrange the process steps in the first and second level of detail.

Furthermore the process model provides the possibility to arrange the process steps of the third level of detail according to the development task. This is done by the role of operational planning. In contrast to the first levels, the third level of detail has no predefined order of process steps given or recommended in general. This implies that the order of process steps is different for different parts and sub-assemblies of the product developed. Thus the process model can be applied recursively in a development project for the whole system, partial systems and single parts of the product.

4.3.3 Consideration of the Design Situation for Process Planning

The design situation is an important part of the approach towards situation specific process planning. In literature there are many different approaches that describe the design situation concerning different objectives (*e.g.* Badke-Schaub and Frankenberger 2004; Hales and Gooch 2004) that illustrate the importance of the situational influence on engineering work. As these descriptions of the design situation are either very extensive or have a different focus, a description of the design situation applicable for this approach was developed.

The situation description applied in this contribution takes into account the different levels of process planning. That means that different parameters are used to describe the situation on the two levels.

The interaction between these levels of planning, *i.e.* providing the possibility to switch between both levels of detail while planning, is essential in order to gain a high value of information and transparency of the process plans. To achieve a context description that is feasible for process planning, the approach focuses on factors that can be measured or estimated at the beginning of a process. Some of the parameters are only relevant for one of the levels of decomposition while others apply for both levels (Roelofsen *et al.*, 2007a).

The general parameters for the situation description chosen in this approach are: design problem and requirements, process results (*i.e.* the required outputs), degree of novelty and complexity of the task.

The situation-describing parameters that are considered on the project level are: customer, risk, project constraints, structure of the design problem and the number of units produced.

On the operational level the influencing factors regarded are: product models available/process up to the present point (input), required output/planned succeeding process steps, structure of partial design problem, operational constraints (organisational, individual, environmental prerequisites) and interdependency with other process participants (number of interfaces).

This situation description covers the most important parameters that influence the design process. The parameters were chosen in collaboration with the partner companies.

The description of the design situation on project level is used to recommend the different process steps to the role "process planning". For example for a high risk project different steps are recommended than for a low risk project. To these extra steps, which are recommended for high risk projects, belongs amongst others the creation of a functional model of the product. Another example is the number of units produced. If only one unit is produced, the use of hardware prototypes is not recommended for the whole system, whereas it is recommended *e.g.* for mass production.

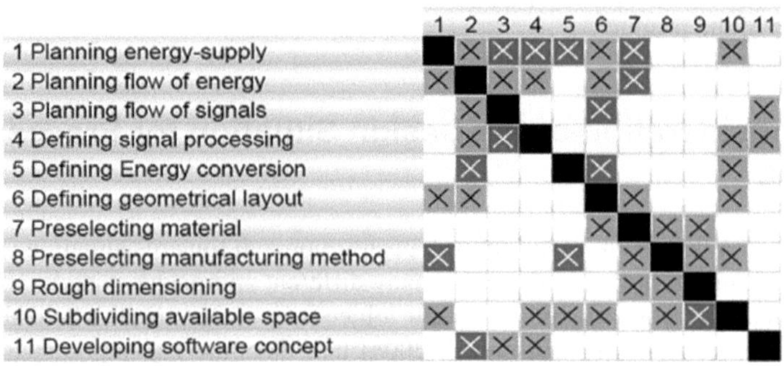

Figure 4.3. DSM of the process steps of the third level of detail (Roelofsen *et al.*, 2008)

Process planning on the operational level will be described applying part of the FORFLOW Process Model. The exemplary sub-steps on the third level of process decomposition are: "Planning energy-supply", "Planning flow of energy", "Planning flow of signals", "Defining signal processing", "Defining energy conversion", "Defining geometrical layout", "Preselecting material", "Preselecting manufacturing method", "Rough dimensioning", "Subdividing available space" and "Developing software concept" (see Figure 4.3; the numbers of the process steps are randomly assigned by the tool used to analyse the process matrix).

The sub-steps have to be arranged in a way that large iterations are prevented and small iterations are supported. This is done by a DSM. The approach is based on the use of time-based DSM as described by Browning (2001). In this DSM the influence of the sub-steps on each other is represented according to the development project on hand. Relations are defined as the dependency of one process step on the results of another step. The project on hand is classified by the design situation parameters. This project classification is used to address different kinds of project situations according to which the DSM is filled differently (Roelofsen *et al.*, 2008).

4.3.4 The Procedural Model

The procedural model used for the approach towards situation specific process planning bases on the Product Model Driven Development (PMDD) approach introduced by (Roelofsen *et al.*, 2007b). The procedural model introduced then was complemented on the project level by a "Project Process Plan" and the "Project Targets" as can be seen in Figure 4.4. In this context product models are defined as all artefacts that contain information about the product that is under development.

According to this model, product development starts with a request by an internal or external customer, which initiates a project. Before work can be started at operational level, project planning has to fulfil some tasks. First, the project situation is analysed to recommend process steps and afterwards the tailoring of the process model for the first and second level of decomposition is done. The result is the overall "Project Process Plan" that is the referential plan throughout the project. This plan is only adapted if major changes occur in the design situation on project level. Otherwise, the project plan and the project targets derived from this plan, as well as the customer request are necessary to keep the project goals in mind and are used to check the fulfilment of targets during the project.

The first step on the operational level is the analysis, which product models are available. After the first customer request these can be documents provided by the customer or the aforementioned project targets, in later project phases product models can be *e.g.* sketches, CAD-data, prototypes and so on. The information about product models available is one of the main inputs for the situation analysis on the operational level. As described above, according to the situation analysis, recommendations for the order of process steps are provided.

After planning the order of process steps, the process is carried out. When a process step is completed, the results are matched with the project goals. A project is completed, if the product models generated in the current process step correspond to the ones defined as the project goals. As long as the models do not match, new operational process steps have to be started. This loop implements the connection between the project and the operational level. Moreover this allows the recursive application of the procedural model for different sub-systems and product parts.

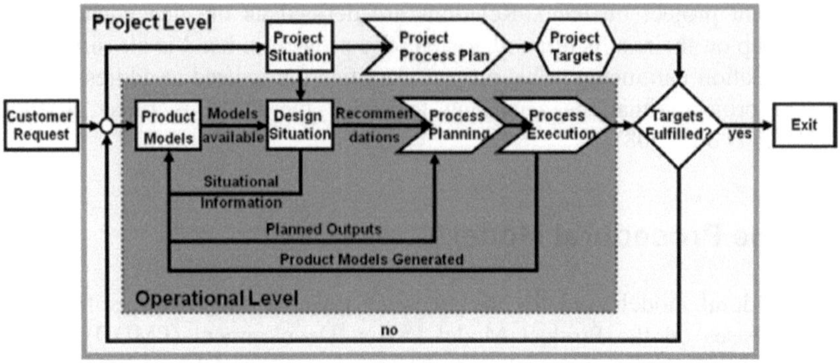

Figure 4.4. Procedural model PMDD

The focus on product models as the main results of development processes is expected to facilitate a development workflow, as it improves process transparency. This is achieved by providing an overview of the product models available and taking into consideration the design situation for every new process step. The link between project and operational level helps to keep the overall project goals in mind for the next process steps (see Roelofsen *et al.*, 2007b for further information).

4.3.5 Integration of Design Situation, Process and Procedural Model

This chapter deals with the integration of the different aspects discussed above. It establishes the relation between the FORFLOW Process Model, the procedural model PMDD and the analysis of the design situation. The approach is implemented in the Process Navigator (PN) that integrates the main results of the research alliance FORFLOW. The PN supports engineers in many different aspects of their daily work, as *e.g.* searching for specific documents, supporting the exchange of different types of CAx-data, connecting the design process and product models *etc.* (Meerkamm *et al.*, 2009). All information about a development project (*i.e.* process information, product models, searches) is stored in a common knowledge base. Here it is important that data is stored only once. If there is *e.g.* a PDM-system in use in the company, there is a risk for data inconsistency. This has to be regarded when the PN is implemented in a company.

Here, the support of planning product development processes is focused. The PN enables the management of design processes and the definition of a flexible workflow that is applicable in product development (Lauer *et al.*, 2008). One of the major drawbacks that result from this flexible workflow is up to now the missing possibility to prevent the engineer from going back to process steps that have to be carried out before certain milestones are reached. To ensure that approved designs cannot be changed after approval is the next optimisation step for the software system.

When a development project is started, the person or group responsible for planning the design process for the overall project starts by defining the project, planning the process on the first and second level of detail, defining milestones and assigning different roles that have to carry out the respective process steps during operational work. All these working steps are carried out in the GUI-view (Graphical User Interface) that is implemented for project management.

On the operational level, there are two different views of the process. One view shows the steps assigned to a specific role, *i.e.* the steps assigned to the current user; the other view shows all process steps assigned to the project. This information is displayed on one part of the PN screen, according to the view chosen by the user. On the other hand there is a variety of different views that gives project specific information to the engineer. Amongst others there is information supplied about the product models required to carry out a process step and whether these models are available or not, as well as the product models that have to be changed or generated during the respective process step.

The flexible process execution is then supported as follows: the process steps assigned to a specific role are displayed to the responsible persons, all process steps can also be viewed by all participants.

The process information consists of three parts that are displayed in both views. The first part is the "Next Step" assigned in the planned process order, either for the role or in the general process, depending on the chosen view. The second part is called "Design Situation". In this part, the surrounding process steps of the current one are displayed. The third part shows all process steps (on the first level of detail of the process model). The engineer working on the operational level is free to jump between process steps. That means he does not have to follow strictly the process order that was predefined on the project level but can adapt the process to his needs. The process steps displayed in the "Design Situation" are shown in the order derived from the process DSM. The information about all process steps is given in order to show the importance of the process step at hand for the entire project and to give the opportunity to switch to steps that are further away (in terms of the order of process steps) from the current task than the ones displayed in the "Design Situation". Thus it is possible to roughly plan the design process on the project level and give enough room and support in process execution on the operational level to enable fast reaction to changes in the design situation. The person working on the operational level gets extensive information concerning the planned process order, interfaces of the steps assigned to his role and other steps as well as the "big picture" of the development project.

Another advantage that can be gained by applying this approach to product development projects is that the process order as it is carried out is documented by the PN, so that successful processes can be reused in follow up projects. Thus a more effective usage of "lessons learned" is implemented. This way it is not always necessary to run the analysis of the process-DSM every time, but process orders can be reused if a similar situation occurs and successful processes are documented historically.

4.3.6 Evaluation

The approach as it is implemented in the Process Navigator is evaluated in two different ways. On the one hand a pilot project was started at one of the industrial partners, on the other hand the functionality of the PN was tested during interviews and workshops. The evaluation was carried out by different groups of industrial practitioners as well as students and researchers. The industrial partners of the research alliance had the possibility to test the PN during a workshop and give their feedback. In that workshop about 30 persons used the software simultaneously. Moreover there were workshops held at five of the industrial partners, where the PN was tested in more detail. In addition to that three Formula Student racing teams tested and evaluated the Navigator concerning their needs. During these workshops the participants had to work through a small development project with assigned roles and tasks. After applying the PN to this small project the participants were asked for their feedback, concerning functionality, ergonomics and practical applicability of the software.

In addition to this the application of the process planning methods introduced in this contribution is planned for a students′ project in order to identify potential for further method improvements.

The general feedback regarding the PN was very positive. The different types of support provided for both project planning and operational planning were rated very useful. Still there is the aforementioned problem that already existing software could lead to inconsistency of information. Moreover the acceptance of yet another software tool by the engineers was questioned. To achieve the necessary acceptance for the Process Navigator it is necessary to clearly show the engineers the benefits of applying the Process Navigator in their daily work.

The pilot project at the partner company is still under way and thus a detailed review of the evaluation results concerning the pilot project is not possible at this point. For further information about the evaluation of the PN please refer to (Meerkamm *et al.*, 2009)

During a workshop with industrial participants, challenges in development process management were discussed. One of the challenges often referred to was that process participants often do not want to apply predefined processes or do not feel responsible for the outcome of these processes. Moreover different "types" of process participants were described: on the one hand those that feel that processes are a restriction to their work; on the other hand those that explicitly want to be lead through the processes they are participating in. All in all the acceptance of the processes defined in a company was discussed thoroughly. This acceptance is hoped to be increased by the application of the approach introduced in this paper. The possibility of either following the tailored process or switching between process steps as it seems fit for operational working supports both "types" of process participants as they were described during the workshop. Moreover the direct influence on the process shall increase the acceptance of the processes and give deeper insight into the importance of the respective processes.

4.4 Conclusion and Future Work

The approach towards situation specific planning of development processes introduced in this contribution brings together a number of approaches that support situation specific planning of design processes that were developed in the research alliance FORFLOW. Combining these approaches increases their applicability in practice and improves the support that the approaches provide separately. In the Process Navigator the partial approaches displayed in this contribution (the FORFLOW Process Model, PMDD, the description of the design situation, the methodical support in choosing and arranging process steps) are implemented. This implementation has been evaluated both in academia and industry, which lead to positive results.

As the process planning approach itself is part of the Process Navigator and was not yet evaluated separately, the evaluation of this approach is still to come. For this purpose, different development projects that are carried out by students will be analysed concerning their process planning. Based on this analysis students will apply the framework for planning a development project. Moreover, the application at one of the industrial partners of the research project is taking place at the moment, in order to prove the practical applicability of the approaches presented here.

4.5 Acknowledgements

The results presented in this contribution were generated in the research alliance FORFLOW financed by the Bayerische Forschungsstiftung.

4.6 References

Badke-Schaub P, Frankenberger E (2004) Management kritischer situationen. Springer, Berlin, Germany

Browning T (2001) Applying the design structure matrix to system decomposition and integration problems: A review and new directions. IEEE Transactions on Engineering management, 47(3)

Hales C, Gooch S (2004) Managing engineering design. Springer, London

Krehmer H, Eckstein R, Lauer W, Roelofsen J, Stöber C, Troll A *et al.* (2009) Coping with multidisciplinary product development - a process model approach. In: Proceedings of the International Conference on Engineering Design (ICED'09), Stanford, CA, US

Kerzner H (2006) Project management. Wiley, NY, US

Lauer W, Faerber M, Roelofsen J, Jochaud F, Jablonski S, Lindemann U (2008) Process management system for the integration of situation dependent process planning. In: Proceedings of the 2008 IEEE Conference on Industrial Engineering and Engineering Management (IEEM08), Singapore

Lindemann U (2007) Methodische entwicklung technischer produkte. Springer, Berlin, Germany

Maffin D (1998) Engineering design models: Context, theory and practice. Journal of Engineering Design, 9:(1)

Meerkamm H, Henrich A, Jablonski S, Krcmar H, Lindemann U, Rieg F (2009) Flexible prozessunterstützung in der produktentwicklung: Prozesse - daten - navigation. Shaker, Aachen, Germany

Pesic M, van der Aalst WMP (2006) A declarative approach for flexible business processes management. In: J Eder, S Dustdar *et al.* (eds.) Business Process Management Workshops. Springer, Berlin/Heidelberg, pp 169-180

Roelofsen J, Baumberger C, Lindemann U (2007a) An approach towards situation specific planning of design processes. In: Proceedings of the International Conference on Engineering Design (ICED'07), Paris, France, pp 193-194

Roelofsen J, Lauer W, Lindemann U (2007b) Product model driven development. In: Proceedings of the European Concurrent Engineering Conference (ECEC'07), Delft, The Netherlands, pp 20-26

Roelofsen J, Krehmer H, Lindemann U, Meerkamm H (2008) Using the design-structure-matrix for the avoidance of unnecessary iterations. In: Proceedings of the 10th international DSM Conference (DSM'08), Stockholm, Sweden, pp 209-211

Vajna S (2005) Workflow for design. In: Clarkson PJ, Eckert CM (eds.) Design process improvement. Springer, London, UK

VDI-Gesellschaft Entwicklung Konstruktion Vertrieb (1993) Systematic approach to the development and design of technical systems and products. Verein Deutscher Ingenieure, Düsseldorf, Germany

Wynn D, Eckert CM, Clarkson PJ (2007) Modelling iteration in engineering design. In: Proceedings of the 16th International Conference on Engineering Design (ICED'07), Paris, France

Chapter 5

Process Model Based Methodology for Impact Analysis of New Design Methods

R. Koppe, S. Häusler, F. Poppen and A. Hahn

5.1 Introduction

What is the impact of new design methodologies and methods on the industrial product development process and its productivity? This is one of the key questions for today's industrial engineering companies when they decide on the introduction of new methods or tools, as well as for researchers and tool providers when they assess their ideas and work. A quantitative (at best monetary) assessment of a new design method's value and impact will increase a manager's decision basis significantly and will lead to well directed investments in process optimisations.

It is not sufficient to analyse the impact of a new design method just for the specific process step it is applied to. Implications on following steps, sometimes on the whole engineering process must be considered as well. Furthermore, decisions on the use of new design methods must be made on a multi-criteria basis through the well-known cost/time/quality triangle (*e.g.* Burghardt, 2006).

A pilot project case study using the new design method would be a possibility. Although such a study would deliver some insight concerning the value of new design methods, it unfortunately consumes money and human resources. Furthermore, one project only allows limited control and therefore does not deliver statistically significant results (Münch *et al.*, 2005). In another context with different parameters like a different product, different designers, or different activities, the evaluation results can vary. Therefore, a deep understanding of the engineering process and its inherent cause-effect relationships is needed to allow a reliable impact estimation of a new design method or tool. The contribution of this paper is a tool-supported methodology to set up and perform change impact analyses for new methods and tools on an existing industrial development process using process and product (related) models.

The remainder of this paper is structured as follows: Section 5.2 summarises related work regarding the investigation of cause-effect relationships in product development processes and impact analysis of process changes. Section 5.3 describes our methodology with an extract of a hardware design process and a process change introducing a new development tool. The last section concludes the paper and gives an outlook on future work.

5.2 Related Work

Related work on impact analysis mainly focuses on applications in Software Engineering. In contrast, there are far less applications in other development domains (like Hassine and Barke, 2008). The main problems are about the same though. This is why we mainly refer to established work in Software Engineering.

In Software Engineering multiple research on Software Process Simulation (SPS) exists that was initially introduced to Software Engineering by Abdel-Hamid and Madnick (1991). SPS is used (amongst others) by software project management to estimate project-level performance, carry out 'what-if analyses', and predict the impact of process changes. For an overview on conducted work of the past two decades, the authors refer to Muller and Pfahl (2008) and Zhang *et al.* (2008a). In general, these models are calibrated to represent the real world behaviour of development processes through empirically gained data (Raffo *et al.*, 1999). This leads to the area of empirical Software Engineering that tries to understand the software development process and its cause-effect relationships. The (possible) impact of a cause is proportionally related to the level of detail, objectivity and precision that is required to model and analyse the cause-effect relationships adequately. In this area the idea of Software Engineering Laboratories (SEL) emerged in academia. SELs are environments to perform Software Engineering related experiments in a specific domain and for a specific purpose (see *e.g.* Rombach *et al.,*1993; Rombach, 1999). Münch *et al.* (2005) mention that SELs did not get a broader acceptance in practice as experiments are expensive and bear the risk of failure. Therefore, they propose the Virtual Software Engineering Laboratory (VSEL) concept that incorporates reusable SPS modules as an addition to traditional experiments. Simulation models could be used as directed guidance as to where further empirical research is advised. For an overview on experiments in empirical Software Engineering, we refer to the surveys published by Sjoeberg *et al.* (2005) and Hannay *et al.* (2007).

Current process impact analysis and estimation work abstracts from detailed product information or focuses on specialised topics of process changes. With our approach we combine the usually separated product model with the development process model itself.

5.3 Methodology

The aim of this methodology is to provide guidance for a stepwise concretisation of studies on development process change impact. The result is in an evaluation plan describing necessary activities for process modelling, data collection, experiments, data analyses and impact estimation.

The methodology applies a probability based process model to describe process elements, their relevant properties and cause-effect relationships between process elements and properties. Such cause-effect relationships are the foundation to study the effects of a new method on the process behaviour as for example possibilities of re-iterations and activity effort. We use process simulation which provides the potential to compare an as-is process to different process alternatives that implement new methods.

The following sections will detail the methodology to perform impact analysis of method and tool changes on an existing development process. The methodology can be divided into two major parts. Section 5.3.1 describes the first part on modelling the as-is process and its cause-effect relationship whereas Section 5.3.2 describes the impact analysis of a new development method/tool.

We will illustrate our approach underpinned by a simplified design flow for an Application Specific Integrated Circuit (ASIC), derived from one of our development processes. In this development process we are interested in the introduction of a new tool and its impact on the process effort, as well as its influence on product characteristics (chip quality).

5.3.1 Modelling the Existing Development Process

One of the main challenges in change impact analysis is to understand and model the as-is development process and its cause-effect relationships. This as-is development process forms the reference for the evaluation of the impact of changes. The next subsections will detail each of these steps following Rus *et al.* (2003):

1. define impact analysis goals (5.3.1.1);
2. describe static process model (5.3.1.2);
3. identify and describe cause-effect relationships between process and product model entities/parameters (5.3.1.3);
4. collect and analyse empirical data to quantify cause-effect relations (5.3.1.4);
5. quantify cause-effect relationships (5.3.1.5).

These steps will lead to a process description that enables a simulative impact analysis of changes.

5.3.1.1 Define Impact Analysis Goals

It would be a hopeless undertaking to create an all-embracing process model of even a reasonable complex development process. So in a first step analysis goals are defined to narrow the scope of the impact analysis. Thereby, we reduce the effort for modelling process elements and their properties to a selective required minimum. For this purpose the goal-question-metric (GQM) approach is used to refine goals into questions. These are needed to decide on the relevant parameters and metrics like *e.g.* the measurable values effort and costs of activities.

As an example this paper follows the task to estimate the impact of a new tool that generates power optimised Register Transfer Level (RTL) code from a SystemC description (compare Figure 5.1 and 5.5). Previously, this transformation process was done manually. The following questions are of interest:

- Does the tool reduce overall design effort?
- Is there a quality gain?
- Does it reduce the risk for costly iterations in the design process?

The main influence to be analysed is the change in design effort from *RTL Design* only to the combined effort for the new *Design Transformation* from SystemC to Very High Speed Integrated Circuit Hardware Description Language (VHDL) and the remaining *RTL Design* (possible iterations) illustrated in Figure 5.5. Furthermore, the impact on physical characteristics of the chip like its power dissipation, timing (clock frequency) and silicon area (size), as well as on its quality characteristics like *functional correctness* and *maintainability* of the VHDL output have to be considered. For simplicity reasons, we focus on *functional correctness* and *maintainability*.

5.3.1.2 Describe Static Process Model

We firstly have to understand and document the development process to be modelled. These representations lead to a first refinement of the problem to be analysed (formulated in Impact Analysis Goal Definition). Process documents, as well as interviews with people involved in the process, serve as sources of information for this activity. The process model shows which activities transform which artefacts and how information flows through the model. Classical elements in a process description are the following:

- activities, tasks and their interdependencies;
- milestones and their needed results (decision gates);
- work products: description of design artefacts as input and output of defined activities, tasks and milestones;
- resources and their properties: tools to be used in specific activities with associated cost models, as well as actors, their cost models and their skills (when to be considered in an impact analysis).

These general elements can all be described by existing process modelling languages. As a basis for our process meta-model, we propose to use the OMG standard Software & Systems Process Engineering Meta-Model (SPEM 2.0) (Object Management Group, 2008). In this we do not restrict ourselves to any proprietary process modelling language.

SPEM 2.0 allows the description of engineering processes and associated process elements like activities, transitions, actors and tools. Of course, this is not sufficient for simulation based impact analysis of new methods, as it is just a descriptive process modelling language.

We agree with the authors Deissenboeck and Pizka (2007) that full empirical studies of process changes are in most cases not feasible. Therefore, we follow their idea and use a probabilistic process model that precisely describes probabilities of activity re-iterations. Such transition probabilities can depend on modelled properties, value ranges and value progress, as well as uncertainty conditions over the dimension of time.

According to this, we enhance the process model with concepts to describe:

- quality and size characteristics of work products;
- effort models for activities;
- iteration possibilities between activities with associated probability (can be static or dynamic).

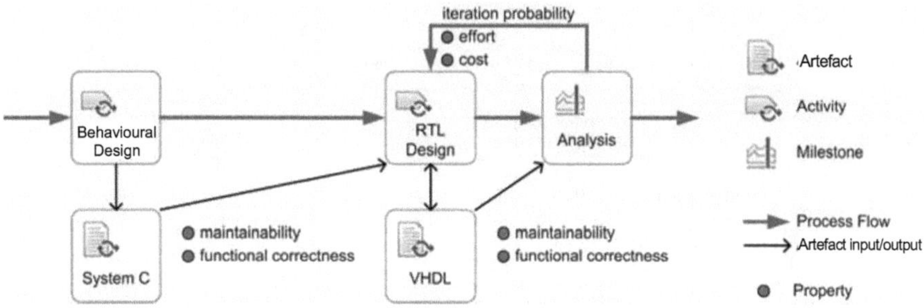

Figure 5.1. Detail of the modelled as-is process with relevant activities, artefacts and properties, to be compared to the changed process in Figure 5.5

All mentioned elements are modelled using our IMPACT process modelling tool. Figure 5.1 shows the modelled as-is process which consists of the activities *Behavioural Design*, *RTL Design* and *Analysis* connected by transitions (blue arrows). The *Analysis* activity verifies *functional correctness* of the input artefact *VDHL* using *e.g.* defined test cases. If *functional correctness* is not verifiable, an iteration to *RTL Design* activity is necessary. The *iteration probability* describes the risk for such iterations and depends on various properties (see next section). The result is a static process description as a baseline to describe cause-effect relationship.

5.3.1.3 Identify and Describe Cause-Effect Relationships Between Process and Product Model Entities/Parameters

Based on the previous step, we have to enrich the static process model with adequate cause-effect relationships. A first qualitative description of these relationships can be done through cause-effect diagrams (Rus *et al.*, 2003).

Figure 5.2 shows parameters with possible influence on an activity's success/iteration probability as mentioned in 5.3.1.2. The probability describes the likelihood to finish an activity successfully and proceed with the next step or iterate. Obviously, the iteration risk directly depends on this *Success Probability*. It is also thinkable that there is more than one possible transition after "failed". In this case the *Success Probability* or better the fail probability has to be distributed along all the alternative paths. This in turn depends on the fulfilment of defined goals (*Goal Fulfilment*) influenced by:

- *Goals* defining the targets for a specific quality characteristic like power, timing, area or functional correctness.
- Characteristics of the verification method in use, *e.g.* accuracy. The better the accuracy of a specific verification method, the higher the probability to find possible flaws in the design artefact. Designer abilities influence quality and certainty of a verification (*Verification Quality and Certainty*).
- The properties of an activity's output design artefact (*Output Artefact Property Probability Distribution*) is modelled as a probability distribution. The probability to produce an artefact with a certain property value is influenced by input properties like characteristics of *Input Artefacts*, used tools and involved actors.

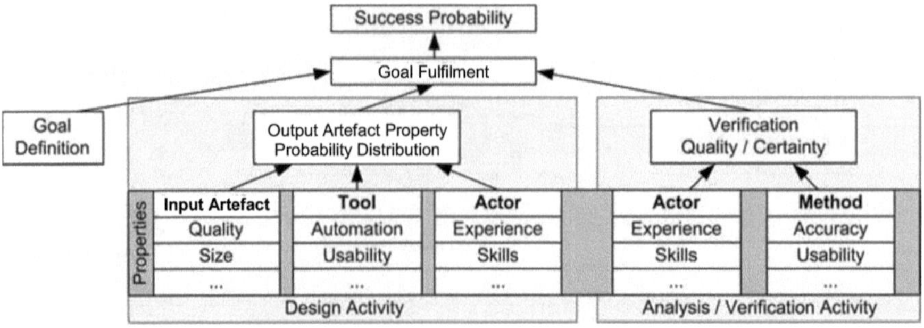

Figure 5.2. Possible impacts on success probability of an activity

To reduce modelling complexity, cause-effect relationships are modelled stepwise, activity by activity. To demonstrate this, Figure 5.3 shows a detail of our process which we decided is of interest for the process change:

- higher *functional correctness* has a positive impact on *Analysis* success,
- higher *maintainability* of VHDL code decreases effort during re-iterations of *RTL Design* and
- SystemC *functional correctness* has a positive impact on VHDL *functional correctness*.

Figure 5.3 shows qualitative cause-effect relationships (dashed arrows), expressed positive "+" impacts (better maintainability "+" results in reduced effort "+", positive effect). Where possible and practical, such cause-effect relationships should be concretised through weighting (*e.g.* more positive, positive, no, negative, more negative impact). This allows for a more accurate modelling of tendencies for quantifications, as well as impact analysis and estimations.

Qualitative or semi-quantitative simulation methods can be used as mentioned for example in Kuipers *et al.* (1988), Takáč (1997) and Zhang *et al.* (2008b) to describe qualitative cause-effect relations. The first simulations with sensitivity analyses can already be used to explore the spectrum, weighting and relation of properties and their potential impact. In the next step quantitative data will be collected and analysed to detail the qualitative described models from this section.

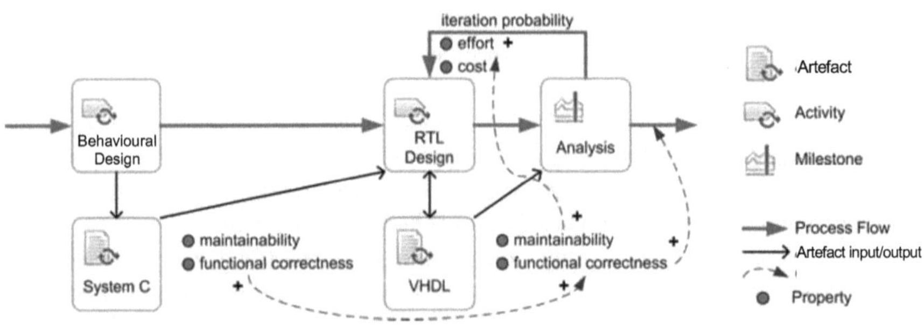

Figure 5.3. Change relevant as-is detail with cause-effect relationships (dashed arrows)

5.3.1.4 Collection and Analysis of Empirical Data to Quantify Cause-Effect Relationships

Qualitative or quantitative statements describe and characterise process elements and their properties, as well as the relationships and cause-effect relationships through mathematical models. These models (*e.g.* for the probability of iterations and effort) will be used to simulate a process and its specific behaviour with the aim to do impact analyses and impact estimations.

We use two approaches to gather data which describes a process and its inherent process characteristics:

- **Extraction of expert knowledge** through focused interviews, observation or manual logs. Such descriptions could be qualitative, as mentioned in step 5.3.1.3, or quantitative, as well as semi-quantitative which describes an aggregated estimation. Further interviews will be used to validate modelled process characteristics against real world experience in the specific context.
- **Automatic data collection** is used to gather product and process data from industrial engineering processes as *e.g.* by analysing log and report files or from user interaction tracking. We have implemented several tools to support this, *i.e.* in Eclipse workbenches and plugins. This data collection infrastructure can be used in projects, field studies, as well as experiments.

Example: Interviews of experts suggest that maintainability is dependent on McCabe's Cyclomatic Number, nesting level and information flow of the VHDL (Mastretti *et al.*, 1995). The automatic data collection is configured in such a way that the relevant parameters are extracted from the process during the flow and compute a measure for the VHDL's maintainability.

Validity and significance of the collected data are important factors for suitable and accurate process and behaviour models. Furthermore, impact analyses, sensibility analyses and impact estimations depend on the gathered and aggregated data. For details on data collection methods and data validity we refer to Yin (1994), Creswell (2002), Bryman (2008).

5.3.1.5 Quantify Cause-Effect Relationships

The intuitive, qualitative cause-effect relationships modelled in step 5.3.1.3 have to be verified, concretised or even corrected if indicated by the quantitatively gathered data of step 5.3.1.4. This is done according to the specific domain and company situation and the specified goals of step 5.3.1.1 to avoid unnecessary effort and to only focus on what is necessary for the analysis.

We create the quantitative models through data analysis methods (*e.g.* sensitivity analyses, correlations and regressions). The representativeness of measures, as well as their distribution and their relation to measures of other properties have to be considered carefully to eliminate incoherence (*e.g.* Raffo, 1993; Raffo, 1999).

Quite often a range of values or probability is more likely the result for gathered data. Looking at *e.g.* the project dimensions effort, cost and quality, we can then describe these by means of probability distributions. Such distributions are able to describe more or less likely values for a property according to other process characteristics as mentioned in 5.3.1.3.

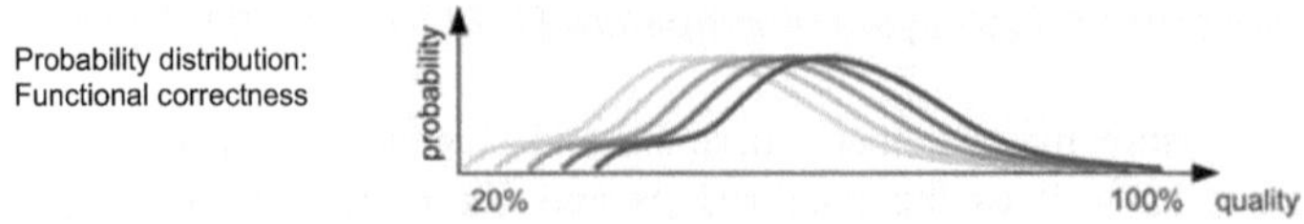

Figure 5.4. Probability distribution for the quality of functional correctness for multiple iterations of an activity

For our example the probability distribution of the property quality (*functional correctness*) is shown in Figure 5.4. The distribution is modelled by experts' experience and/or gathered data. On the x axis, the quality value is shown. The height of the curve (y axis) stands for the likelihood of the corresponding quality value to be reached. With increasing iterations the distribution of possible values should move from lower quality to higher quality values (called added value). It is also possible that the shape of the distribution function changes during this progress.

At the end of the as-is process modelling activity the modelled process with its behaviour shall be simulated to calibrate and validate the modelled as-is process against the real world experience in the organisation. Surely, this is an ongoing process to improve and keep the alignment of the model with the real world.

Our IMPACT tool provides a process library where characterisations and patterns of activities, roles, tools and cause-effect relationships can be stored and reused to reduce the initial modelling effort. Also imports from existing process descriptions are possible.

5.3.2 Impact Analysis

The impact analysis of process changes consists of three steps as described in the following.

1. description of the new method or tool (5.3.2.1);
2. simulation of the as-is process against the changed process as a virtual experiment for impact analysis (5.3.2.2);
3. validation and improvement of simulation models (5.3.2.3).

5.3.2.1 Description of the New Method or Tool

The as-is process presents the foundation for cause-effect relationships and process behaviour where the new method shall be integrated into. So the process description around the new method or tool is the next step in an impact analysis.

It is the objective to describe the differences in the new method or tool according to process elements and relations, described in the as-is process. The following aspects have to be considered when modifying the specified as-is process by a new method or tool characterisation:

- new activities include additional inputs and outputs;
- new activities have specific effort models;
- as-is process activities become obsolete;
- the new method can have different cause-effect relationships that were not considered in the as-is process, *e.g.* how artefacts and their properties are influenced by the new method or tool and its characteristics.

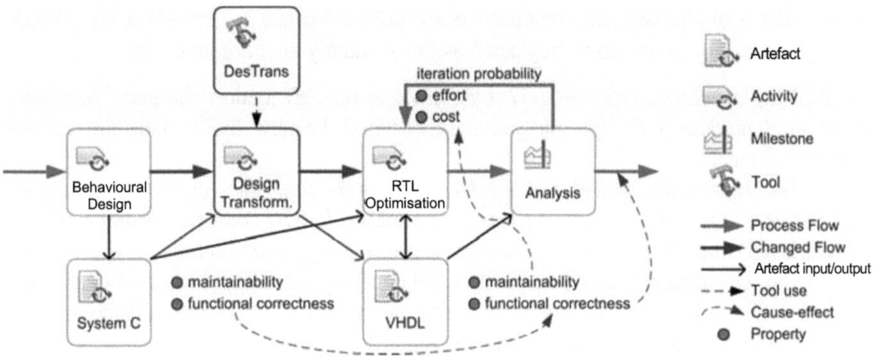

Figure 5.5. Changed process with its cause-effect relationships

Figure 5.5 shows the changed process (red arrows and process elements) characterised through the introduction of the new activity *Design Transformation* which uses the tool *DesTrans* to synthesise *SystemC* to *VHDL* code. For this process change, we have to model the usage and tool characteristics of *DesTrans* which influences the quality of the *VHDL* code (*e.g. functional correctness* and *maintainability*) and other artefact properties.

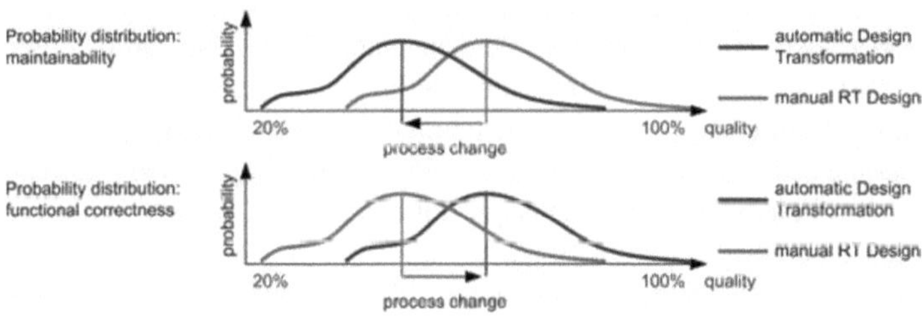

Figure 5.6. Change effect impacted by properties (*e.g.* tool automation level and actor skills)

Figure 5.6 shows the potential qualitative effect of the process change on the properties' *maintainability* and *functional correctness*. We made assumptions that we have to investigate during the next steps of simulation and impact analysis:

- the new method needs additional design steps in *Behavioural Design* to enrich the *SystemC* model with further information needed for synthesis which increases the effort of *Behavioural Design*;
- the probability of correct functional *VHDL* code correlates with the *functional correctness* of the *SystemC* design as the tool provider guarantees the transformation of correct *SystemC* code into correct *VHDL* code;
- the probability to get high *VHDL maintainability* decreases as a result of automatic transformation and machine generated *VHDL* code;
- RTL Design changes to optional *RTL Optimisation* to optimise e.g. power if necessary. This effort is influenced by the maintainability of the *VHDL* code, but should be significantly lower than the effort of *RTL Design* in the as-is process as a result of automatic *Design Transformation*;

- the tool has specific purchase costs and/or licence costs; effort for *Design Transformation* can be neglected as this is mainly an automatic step.

Will the decreased effort of *RTL Optimisation* pay off against the extra additional effort in *Behavioural Design* and purchase costs of the new tool? There are various possible scenarios:

A) The extra effort in *Behavioural Design* plus the tool costs outweighs decreased effort in *RTL Optimisation* due to less maintainable *VHDL* code (if manual work on VHDL is necessary to optimise e.g. power usage). The new method fails.

B) The automatically generated code is functionally correct and requires hardly any manual work so that a worse maintainability has no severe influence and the positive effect of effort saved for manual transformation is dominant.

Both scenarios are conceivable and simulation should be able to give the answer.

5.3.2.2 Simulation of the As-Is Process Against the Changed Process as a Virtual Experiment for Impact Analysis

The models from step 5.3.2.1 of the new method or tool are integrated in the modelled as-is process. This changed process describes therefore an alternative process. We use simulation as a tool for the following purposes:

- The stepwise execution of an altered process model is a useful function to inspect the behaviour of the change in detail. Time, effort, defined goals, properties of artefacts, process elements and added value can be observed step by step for a process in comparison to alternative processes.
- Multiple complete runs of an altered process cover an expanded view of many possible scenarios (*e.g.* Monte Carlo simulation). Aggregated information from statistical analyses allows mathematical comparison, as well as tracking related analyses for process alternatives.
- Powerful analyses during simulation allows exploration of differences and uncertain parameters for the defined models to estimate best-case, worst-case and more likely scenarios.

The simulation gives an overview of possible behaviours of alternative processes for project dimensions like time, effort and an overview on artefact characteristics and their progress over time.

According to the goals specified in section 5.3.1.1 process changes will be estimated. Besides the well-known statistical methods like minimum, maximum, arithmetic average and median, tools like variance and regression analyses are used to extract useful information from simulation runs. Graphical box plots, histograms, portfolios and value progress over time are usable aids.

The comparison of such aggregated and refined process information enables statements about the difference between process alternatives with corresponding tendencies and best-case, as well as worst-case descriptions. The economic benefit is described as the effort/cost distance between the as-is process and a process alternative including necessary investments for a change, like costs of purchase for tools. Factors for parameters, properties and scenarios can be introduced to take aspects of uncertainty into account. Hence a broad overview of possible and more likely scenarios is given which supports strategic decision making.

5.3.2.3 Validation and Improvement of Simulation Models

The simulation results and their underlying assumptions have to be validated and quantified through aimed (quasi-) experiments. Validation and quantification allows reducing uncertainty of the modelled process behaviour and validation of a statement that a new design method or tool pays off. At the start of an investigation, we will just be able to state conditions under which a method will pay off. For example: The new transformation tool only pays off, when the additional effort in Behavioural Design is significantly below the previous effort of RTL Design and the probability of the need to manually optimise the VHDL code is below 10%. Whether these probabilities reflect practice, needs to be justified either by a tool provider who guarantees this, an expert and experiments or application in real projects. The last point will need more effort for data collection and analysis but will give results with least uncertainty.

5.4 Conclusion and Future Work

This paper presents a methodology to enable impact analyses of new methods and tools on existing engineering processes to support managers in their decision whether a new design method or tool should be applied and pays off. Our methodology implements the foundational requirements for impact analyses discussed in Häusler *et al.* (2009).

The foundation for impact estimations are models of the development process and representation of cause-effect relationships between different process elements. We explained the steps to describe current as-is process. After the as-is process is modelled and calibrated to the organisational characteristics, we described our approach for impact analyses based on an extension of SPEM 2.0 with an underlying Markov chain semantic to enable simulation.

The results of multiple simulations can be aggregated, refined as a meaningful decision basis for managers covering best-case and worst-case scenarios and describe more likely scenarios for a picture of holistic process alternatives.

The effort for the methodology depends clearly on the goals for the impact analysis. It is possible to implement first impact estimations based on expert knowledge or existing process data. Parameters of uncertain models and probability distributions can be varied to explore their effects and break-even conditions for the changed process. We can refine first assumptions through data collection and analysis within experiments or real projects to improve models for added value and effort and reduce the uncertainty of impact estimations.

5.5 References

Abdel-Hamid TK, Madnick SE (1991) Software project dynamics: An integrated approach. Prentice Hall, Englewood Cliffs, NJ, US

Bryman A (2008) Social research methods. Oxford University Press, Oxford, UK

Burghardt M (2006) Projektmanagement: Leitfaden für die planung, überwachung und steuerung von entwicklungsprojekten. Publicis Corporate Publishing. Erlangen, Germany

Creswell JW (2002) Research design: Qualitative, quantitative, and mixed methods approaches. Sage Publications Ltd, Thousand Oaks, CA, US

Deissenboeck F, Pizka M (2007) The economic impact of software process variations. Software Process Dynamics and Agility. Springer, Berlin, Germany, pp 259-271
Hannay JE, Sjoberg DIK, Dyba T (2007) A systematic review of theory use in software engineering experiments. IEEE Transactions on Software Engineering 33(2): 87-107
Hassine A, Barke E (2008) On modelling and simulating chip design processes: The RS Model. IEEE International Engineering Management Conference - Europe (IEMC-Europe), Estoril, Portugal
Häusler S, Buschermöhle R, Koppe R, Hahn A (2009) Towards process change impact analysis in industrial engineering. In: Proceedings of the IEEE International Conference on Industrial Engineering and Engineering Management (IEEM 2009), Hong Kong, China, pp 1489-1493
Kuipers B, Berleant D (1988) Using incomplete quantitative knowledge in qualitative reasoning. In: Proceedings of the National Conference on Artificial Intelligence (AAAI-88), Los Altos, CA, US
Mastretti M, Busi ML, Sarvello R, Sturlesi M, Tomasello S (1995) VHDL quality: Synthesizability, complexity and efficiency evaluation. In: Proceedings of the Design Automation Conference (EURO-DAC'95), Brighton, UK, pp 482-487
Muller M, Pfahl D (2008) Simulation methods. Guide to advanced empirical software engineering. Springer, London, UK, pp 117-152
Münch J, Pfahl D, Rus I (2005) Virtual software engineering laboratories in support of trade-off analyses source. Software Quality Control 13(4): 407-428
Object Management Group (2008) Software & Systems Process Engineering Meta-Model Specification 2.0. Available at: http://www.omg.org/spec/spem/2.0/pdf
Raffo DM (1993) Evaluating the impact of process improvements quantitatively using process modelling. In: Proceedings of the Conference of the Centre for Advanced Studies on Collaborative Research (CASCON'93), Toronto, Canada, pp 290-313
Raffo DM (1999) Getting the benefits from software process simulation. In: Proceedings of the International Conference on Software Engineering and Knowledge Engineering (SEKE'99), Kaiserlautern, Germany
Raffo DM, Kaltio T, Partridge D, Phalp K, Ramil, JF (1999) Empirical studies applied to software process models. Empirical Software Engineering 4(4): 353-369
Rombach HD (1999) Experimentation engine for applied research and technology in software engineering. In: Proceedings of NASA's 24th Annual Software Engineering Workshop, Greenbelt, MD, US
Rombach HD, Basili VR, Selby RW (1993) Experimental software engineering issues: Critical assessment and future directions. In: Proceedings of the International Workshop on Experimental Software Engineering Issues: Critical Assessment and Future Directions. Lecture Notes in Computer Science, Springer-Verlag, Berlin, Germany
Rus I, Neu H, Münch J (2003) A systematic methodology for developing discrete event simulation models of software development processes. In: Proceedings of the 4th International Workshop on Software Process Simulation and Modeling, Portland, OR, US
Sjoeberg DIK, Hannay JE, Hansen O, Kampenes VB, Karahasanovic A, Liborg NK *et al.* (2005) A survey of controlled experiments in software engineering. IEEE Transactions on Software Engineering 31(9): 733-753
Takáć M (1997) Fixed point classification method for qualitative simulation. In: Proceedings of the 8th Portuguese Conference on Artificial Intelligence: Progress in Artificial Intelligence. Lecture Notes in Computer Science, Springer-Verlag, Berlin, Germany, pp 255-266
Yin RK (1994) Case study research: Design and methods. Sage Publications. Thousand Oaks, CA, US
Zhang H, Kitchenham B, Pfahl D (2008a) Reflections on 10 Years of software process simulation modelling: A systematic review. In: Proceedings of the International Conference on Software Process (ICSP'08), Leipzig, Germany
Zhang H, Keung J, Kitchenham B, Jeffery R (2008b) Semi-quantitative modelling for managing software development processes. In: Proceedings of the 19th Australian Software Engineering Conference, Perth, Australia, pp 66-75

Chapter 6

Design Process Planning by Management of Milestones and Options with a Design Attainment Model

Y. Nomaguchi, T. Yamamoto and K. Fujita

6.1 Introduction

In order to enhance competitiveness of a product, a design project should perform an innovative design toward its higher quality and performance. However, innovative design involves many risks because of lack of knowledge or experience, *e.g.* to not finish a task in the predicted time. Milestone management is necessary for accommodating such risks. Even if an innovative design is targeted at the beginning, it is possible that a project manager will change the strategy to a conservative design in order to prioritise risk reduction over the challenge. A manager should plan a design process by assessing a risk of the design, and should decide to switch the strategy at the appropriate milestone while checking progress of the design if necessary.

This paper proposes a method of planning collaborative design process for innovative products by managing milestones and the multiple options of the design strategy through assessment of risk for each strategy. The method aims to perform an innovative design or change it to a conservative backup one in order to accommodate various risks. A planning problem is formulated under the assumption that a design project and its process consist of multiple tasks, each of which corresponds to a component of the product. Multiple design options are assumed for each task. A project manager decides the switches of the task strategy by checking the design progress at the milestones. A design attainment model based on a growth curve and fuzzy inference (Nomaguchi *et al.,* 2008) is adopted to predict the design progress for assessing the risk. This research also demonstrates a planning example of a student formula car project (OFRAC, 2009) and discusses the possibility of the proposed method.

6.2 Issues of Design Process Planning Support

6.2.1 Considering Risk and Milestone Management

Because the design process includes various indistinct factors, there are always risks that unexpected events will happen during the design project, and then the planned lead time, the planned design cost and the planned quality should be revised (PMI Standards Committee, 2008). Milestone management is necessary in order to monitor the design progress, and to possibly change the planned design strategy, *e.g.* switch from the innovative design to the conservative one which has a smaller risk than that of the innovative one.

A milestone is a marker of achieving an identifiable target in any task of the project, *e.g.* completion, endorsement or signing of a deliverable, document or a high level review meeting. As for the design process, the attainment target at the regular review meeting corresponds to the milestone. A project manager checks the progress of the design at a design review meeting, and decides whether or not the current design strategy can proceed.

6.2.2 Indistinctness and Vagueness of Design Process

In milestone management, a project manager has to evaluate the planned design process by predicting the progress after each design review, as well as checking the attainment at the design review. That is, any prediction method of the design progress is necessary for rationalising such management.

Regarding the evaluation of the process planning generally, some methods such as PERT (Program Evaluation and Review Technique) and CPM (Critical Path Method) (Moder, 83) have been practically used especially in the management of construction process and production process. These methods assume that respective tasks of a project can be distinctly defined, and that they are executed in a sequential way. The methods are used to quantitatively evaluate the running time and cost of a project and to minimise them under such assumption. In contrast, tasks of a design project cannot be distinctly defined because it includes creative and collaborative activities. Therefore, quantitative evaluation schemes of running time and cost embedded in the above methods are not well suited to design process planning.

The major difference in creative and collaborative activities is that the borders of exact processes are not clear. The design process is progressive, so that its beginning and its goal are not distinct. In design process planning, it is essential to set a target level for the design attainment and to assess whether or not a planned design process can achieve such a level. These issues require modelling vague aspects of the design process such as the designer's ability, communication efficiency, their complicated mutual dependency, *etc.*

6.2.3 Research on Design Process Planning

The focus on explicitly representing task relationships and their arrangement brings some practical approaches of design process planning such as IDEF0 (Marca and McGowan, 1988) and DSM (Design Structure Matrix) (Eppinger *et al.*, 1994). However, their abilities are limited within the qualitative aspects of design process planning.

To quantitatively evaluate a design process plan, a framework for modelling its vague factors, *e.g.* the ability of an individual designer or the collaboration mechanism of designers, is indispensable. Some methods for quantitative prediction of the design process have been proposed in recent years, and they have introduced several mathematic models for describing the vague factors of design process (*e.g.*, Ostergaard and Summers, 2007; Reich and Paz, 2008). Simulation-based approaches are also available to estimate the stochastic nature of the vague factors, such as rework simulation (Cho and Eppinger, 2005) and the Signposting methodology (Clarkson and Hamilton, 2000). These quantitative approaches assume that time and cost are required for the tasks, and their iteration probability and progress can be distinctly described. However, these factors are usually indistinct at the planning phases especially in creative design projects.

6.2.4 Our Approach

While any mathematical model is necessary to predict the design progress, its prediction inevitably contains inaccuracy because of the indistinctness and vagueness of design achievement. The management of milestones and design options should be in conjunction with any appropriate mathematical model. Thus, this research takes the following two approaches as its key features.

- Formulation of design process planning problem considering the management of milestones and task options.
- Risk assessment through prediction of design progress by the design attainment model.

Section 6.2.1 formulates the design process planning problem by considering the milestone management in the design process. Towards the formulation, this research defines task strategy options, each of which is the plan of the timing and the way of substituting the backup task strategy. The planning problem is formulated as selecting the optimal task strategy considering the backup.

Section 6.2.2 describes the design attainment model, which the authors have proposed (Nomaguchi *et al.*, 2008).

6.3 Management of Milestones and Options

6.3.1 Definition of Task and Task Strategy Alternative

This research introduces some concepts on the management of milestones and options in design process planning.

6.3.1.1 Task

A design process is usually performed through collaboration of multiple designers, or sometimes, collaboration of multiple teams. A designer takes charge of a certain part of the design process, that is, a task. A task usually corresponds to a part of the eventual product. Since each part of a product depends on other part(s), each task also depends on other task(s). Therefore, a designer has to carry out a task while ensuring its consistency with other task(s). This paper denotes the task by T_k (k=1, 2, …, K). K is the number of the tasks consisting of the design process.

6.3.1.2 Task Strategy Alternative

A project manager prepares multiple backup strategies for each task in order to mitigate the risk of the design process. We call each of these the task strategy alternative. As shown in Figure 6.1, multiple task strategy alternatives, *e.g.* from innovative to conservative, are prepared for each task. At the planning phase, a project manager selects the optimal task strategy alternative for each task. This paper denotes the task strategy alternative for task k by A^k_i (i=1, 2, …, I_k). I_k is the number of the task strategy alternatives for T_k.

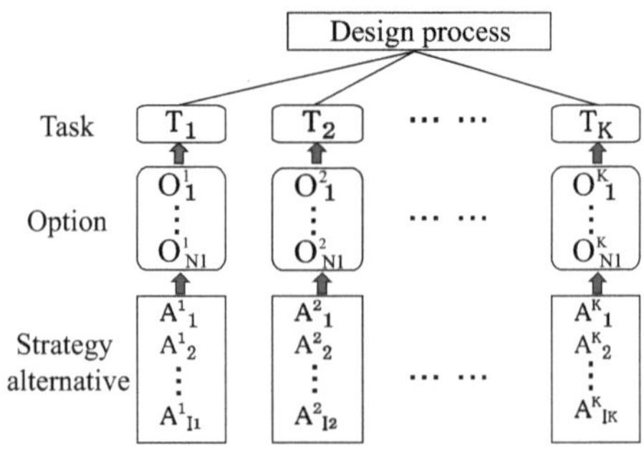

Figure 6.1. Structure of design process

The task strategy alternative is evaluated from the viewpoint of its merit and its risk. The following two parameters are introduced, each of which corresponds to the merit and the risk respectively.

- *Quality potential* (q^k_i) is a relative degree of the product quality yielded when the attainment target of the task strategy is achieved. This corresponds to the merit of the task strategy alternative. In general, the quality potential of the innovative design is higher than that of the conservative design.

- *Task difficulty* (m^k_i) corresponds to the risk of the task strategy alternative. The more difficult the task strategy alternative, the greater risk for achieving the target within the lead time of the design process. In general, the task difficulty of the innovative design is greater than that of the conservative design.

6.3.2 Design Review Meeting

The achievement target at the regular review meeting corresponds to the milestone. A design process manager checks the progress of the design at a design review meeting, and decides whether or not the current design strategy can proceed. This paper denotes the *j*th design review meeting by DR_j, and the design time between DR_{j-1} and DR_j by DT_j. The number of the design review meetings in the design process is denoted by *J*.

6.3.3 Task Strategy Options

Once the milestone management has been devised, the design review meeting is the unique opportunity for switching the task strategy alternative. This research introduces the task strategy option which is the plan of switching or continuing the task strategy alternatives as for all design review meetings. It is denoted by O^k_n (n=1, 2, …, N_k). N_k is the number of the task strategy options for T_k.

The following two categories of the task strategy option are defined.

- *Priority strategy option* - the most innovative task strategy alternative is preferentially performed from the planning to the end of the design project.
- *Backup option* - a more conservative task strategy with less risk which will be switched to if the priority strategy option is aborted at any future design review.

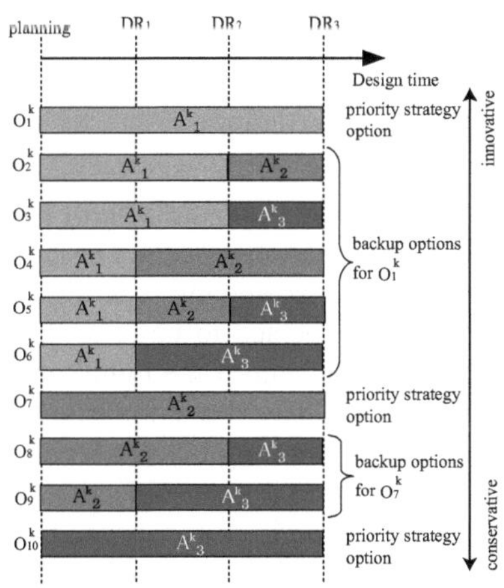

Figure 6.2. Task strategy option

Figure 6.2 illustrates the task strategy options of T_k in the case of $J=3$ and $I_k = 3$. The three task strategy alternatives are sorted in the order of the task difficulty. The task strategy options $O^k{}_1$, $O^k{}_7$, and $O^k{}_{10}$ are the priority strategy options of $A^k{}_1$, $A^k{}_2$, and $A^k{}_3$. The other seven options are the backup options. For example, $O^k{}_2$ is the option that the task strategy $A^k{}_1$ is performed at the beginning, but it is aborted at DR_1, and then the task strategy $A^k{}_2$ is alternatively performed. This is the backup option of $O^k{}_1$.

6.4 Formulation of Design Process Planning Problem

The planning problem performed at DR_j focuses on the decision of the task strategy alternatives performed in DT_j and of all tasks of the design project. The plan is evaluated by a set of objective functions h^k (k=1, 2, ..., K) which quantifies the merit and the risk probability of the selected task strategy of the task T_k, respectively. Because the tasks depend on each other, the value of h^k depends not only on the decision of the task T_k but also on the other tasks. The design process planning problem is formulated as the multi-objective optimisation as follows.

$$
\begin{aligned}
&\text{Find} && \boldsymbol{x} = \{x^1{}_j, \ldots, x^k{}_j \mid 1 \le x^k{}_j \le I_k, x^k{}_j \in I\} \\
&\text{maximise} && h^1(\boldsymbol{x}) \\
& && \vdots \\
&\text{maximise} && h^K(\boldsymbol{x}) \\
&\text{subject to} && \text{given } j
\end{aligned}
\tag{6.1}
$$

where, the design variable $\boldsymbol{x}$ means a set of task strategy alternatives, which form an entire design process plan, as its element $x^k{}_j$ corresponds one for each task.

6.4.1 Progress Prediction by Design Attainment Model

Design progress is quantified by use of a design attainment model for defining the objective function h (Nomaguchi *et al.*, 2008).

6.4.1.1 Growth Curve Model for Design Attainment Prediction

A design process is a knowledge creation process. Design attainment is defined as the level of knowledge that a designer obtains. The higher the design attainment, the higher quality the product is expected to be. The design attainment model is constructed as follows: design attainment of the initial condition, *i.e.* a totally new design in which a designer does not have any knowledge of a task, corresponds to 0, and the design attainment of the limiting highest level, in which a designer has perfect knowledge such as a complete multi-objective optimisation model of a product, corresponds to 1. A numeric value between 0 and 1 expresses an intermediate design attainment that is defined according to the knowledge level of an engineering analysis model, use conditions of the designed product, *etc.* This enables a project manager to set a target level of design attainment.

The following growth curve model is introduced on the premise that in general the design attainment level increases monotonically and continuously.

$$f_i(t) = 1 - (1 - f_{0i})\exp(-m_i \tau_i(t)) \tag{6.2}$$

where, $\tau_k(t)$ denotes the sum of design time spent for task *k* since the beginning of the design process up to time *t*, $f_k(t)$ is the knowledge level of task *i* at time *t*, which is expressed by a numeric value from 0 to 1, f_{0k} is the initial knowledge level that a designer has at the beginning of task *k*, and m_k denotes the difficulty of task *k*. The value of f_{0k} depends on both the designer and the task, while that of m_k depends on the task. Note that a smaller value of the task difficulty parameter *m* means a greater difficulty in this growth curve model.

In collaborative design projects, created knowledge is often revised in order to enhance consistency among tasks, and the knowledge attainment level will thereby decrease. It is empirically rational to assume that the higher the consistency level between tasks, the less the reduction of design achievement, and that the higher the interdependency between tasks, the more meeting time will be required to enhance consistency between tasks. The full lines of Figure 6.3 show an outline of the model of task consistency level g_{kl} between tasks *k* and *l*, which is expressed by a numeric value from 0 to 1 and design revision. This model enables a project manager to predict the design attainment of the task at a certain moment.

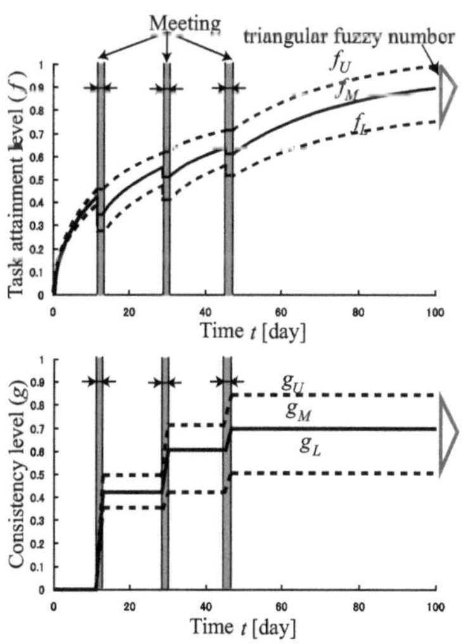

Figure 6.3. Growth curve model for design achievement prediction

6.4.1.2 Representation of Vagueness by Triangular Fuzzy Number

A numeric value of the task attainment of Equation 6.2 contains deflection, because it is a measure of the naturally vague knowledge creation progress. This research adopts a triangular fuzzy number for representing the vague factors. It represents a deflection of vague values as a triangular distribution, while its operations can be done with simple calculations.

Figure 6.3 also shows an example of task attainment with a triangular fuzzy number, f_{kL}, f_{kM} and f_{kU} respectively represent the lower, middle, and upper points of a triangular distribution of the task attainment. The task difficulty m_k is represented by a triangular fuzzy number (m_L, m_M, m_U) so that the task attainment is represented with a triangular distribution. In the same way, Figure 6.3 also shows an example of task consistency with a triangular fuzzy number.

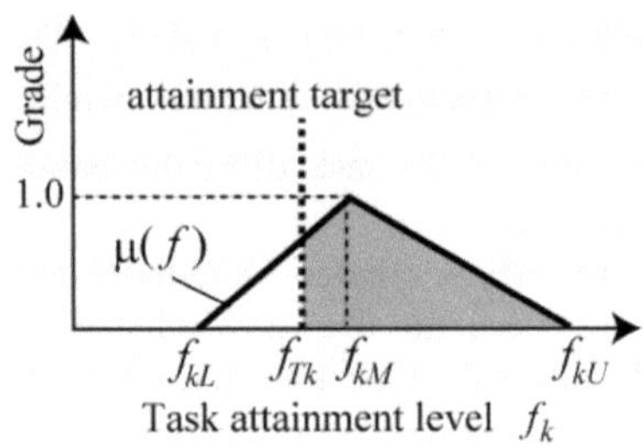

Figure 6.4. Target achievement probability

6.4.2.3 Target Achievement Probability

The target achievement probability p_k of the task k is given as the proportion of the area of a triangular distribution of the task attainment that exceeds the target achievement level f_{Tk} to the area of the entire triangular distribution. When a triangular distribution function is given by $\mu(f)$, p_k is defined with the following equation.

$$p_k = \frac{\int_{f_{Tk}}^{f_{kU}} \mu(f)df}{\int_{f_{kL}}^{f_{Tk}} \mu(f)df} \tag{6.3}$$

Figure 6.4 illustrates the definition of the target achievement probability. The shaded region shows the numerator of Equation 6.3, and the region within the heavy-line triangular shows the denominator of Equation 6.3.

6.4.2 Formulation of Objective Functions

In evaluating the task strategy alternative, the evaluation function h^{*k} for the option $O^k{}_n$ is introduced by summing the target achievement probability and the quality potential.

$$h'^{k}(\boldsymbol{y}) = w_1\, p^{k}(\boldsymbol{y}) + (1 - w_1)\, q^{k}(\boldsymbol{y})$$
$$\boldsymbol{y} = \{y^1, \ldots, y^K \mid 1 \le y^k_j \le N_k, y^k \in \Gamma\} \qquad (6.4)$$
$$0 \le w_1 \le 1$$

Where, $\boldsymbol{y}$ is the design variable which denotes the selected options for all tasks. When the option O^k_n is selected, $y^k = n$. The function $p^k(\boldsymbol{y})$ denotes the target achievement probability of the task T_k under the selected options. The function $q^k(\boldsymbol{y})$ denotes the quality potential of the task T_k under the selected options, that is, the quality potential of the task strategy alternative which is planned to be performed at the final design time DT_J. The parameter w_1 is the weight of the quality potential to the target achievement probability.

As for the evaluation of the task strategy alternative, the existence of a plausible backup option should be considered as well as the evaluation of its priority option. Therefore, the objective function $h^k(\boldsymbol{x})$ is defined by using the value of h'^k of the priority option of the task strategy alternative A^k_i, which is represented by $a(A^k_i)$, and the value h'^k of the backup option of A^k_i , which is represented by $b(A^k_i)$, as:

$$h^k(\boldsymbol{x}) = w_2\, a(A^k_i) + (1 - w_2)\, b(A^k_i) \qquad (6.5)$$
$$0 \le w_2 \le 1$$

The parameter w_2 is the weight of the evaluation of the priority option. Both w_1 and w_2 are set by a project manager such that the purposes of the design process can be reflected in the formulation of the planning problem.

6.5 Formalisation of Design Process Planning Method

The design process planning method is formalised by considering the management of milestones and options. This method consists of three phases as shown in Figure 6.5, *i.e.* (1) planning the design process, (2) carrying out the design process, and (3) checking the design process in the design review meeting, and revising the plan if the design process does not progress as planned.

6.5.1 System for Planning and Evaluating Design Process Plan

In order to verify the possibility of the proposed method, a planning support system is implemented by using Java (JDK 1.5). The system facilitates a project manager to generate a plan of the design process, and predicts the design attainment level of each task under the given plan. The system consists of the five parts as shown in Figure 6.6, *i.e.* a design structure matrix for describing tasks and task dependencies (1), Gantt charts for setting task and meeting schedules (2), lists of task strategy options (3), charts displaying task attainment and task consistency (4), and the calculated values of the objective functions (5).

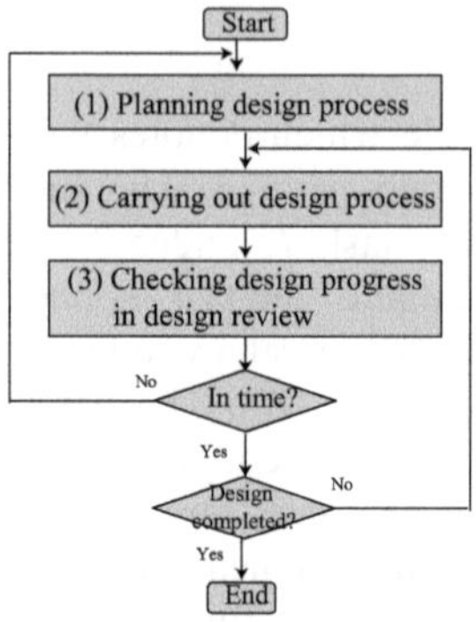

Figure 6.5. Phases of design process planning and management

6.6 Planning Example and Verification

This research uses the student formula project as a case for verifying the method. We interviewed its members and collected the data of the design project performed in 2008. This research made the example planning problem based on the collected data, and applied the proposed method to it. This paper focuses on the verification of the possibility of the objective function h toward management of milestones and options, because it should be the premise of solving the multi-objective optimisation problem noted in Section 6.4.1. As a test case, this section explains the planning of the suspension design, which is one of the tasks of the formula car project, and demonstrates the planning of the task strategy with the proposed objective function.

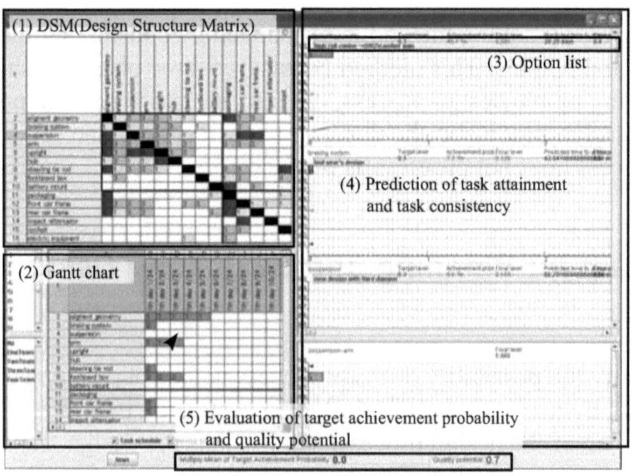

Figure 6.6. A snapshot of planning a formula car design project

For the suspension design task, there are three task strategy alternatives as shown in Table 6.1, *i.e.* alternatives of innovative new design (1 and 2) and an alternative of performing the last year's design (3). The factors for the respective task strategy alternatives *i.e.* the initial attainment level of the responsible designer (f_0), the quality potential (q), the triangular fuzzy number of the task difficulty (m), the attainment target (T), are set as shown in Table 6.1.

The column on the right edge of Table 6.1 shows the values of the evaluation function h' with w_2 = 0.4 for the priority options of the respective task strategy alternatives. The results show that it is optimal to perform the task strategy alternative 2.

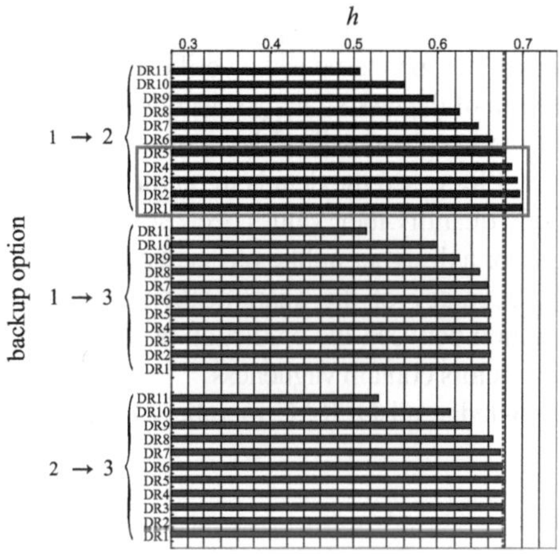

Figure 6.7. Evaluation in consideration of backups

Table 6.1. Data of task alternatives of suspension design

Task strategy alternative	f_0	q	m	T	h'
1. the new design 1	0.1	0.65	0.609,0.891,1.183	0.4	0.3422
2. the new design 2	0.1	0.55	0.775,1.135,1.505	0.3	0.3576
3. the last year's design	0.1	0.4	0.852,1.248,1.656	0.2	0.3200

Figure 6.7 shows the values of the evaluation function h which considers the backup option as well as the priority option. The notation in Figure 6.7 of "1 → 2 DR11" means the consideration of the backup option that the strategy alternative 1 will be switched to the strategy alternative 2 at the 11th design review meeting. The backup options that just one switch takes place are considered in Figure 6.7. Because the project members have 11 design review meetings during the project, there can be 11 backup options for each task strategy alternative corresponding to the switching timings.

The results of Figure 6.7 show that the value of the evaluation function h of the task strategy alternative 1 is greater than that of the task strategy alternative 2 when the backup option of switching the strategy before the fifth design review (as shown inside the heavy line rectangle of Figure 6.7) is considered. Therefore, a project manager can decide that the task strategy alternative 1 has more possibility when the backup option is considered.

6.7 Conclusions

This paper proposed the planning method of collaborative design process for innovative products that manages multiple options of the design strategy and facilitates the change from an innovative design to a backup conservative design. A key feature of the method is the risk assessment of each strategy through progress prediction in conjunction with the management of milestones and design options. Whilst the importance of management of milestones options is well known, it is rarely associated with any appropriate mathematical model. This is a new approach which past and present research in design process planning has not yet achieved. Although the planning example of Section 6 verifies only the proposed objective function, it shows the possibility for the management of milestones and options. One open issue is the verification of the method by solving the multi-objective optimisation of the task strategy selections.

This paper does not discuss the detail of the way of determining the input data, *e.g.* those shown in Table 6.1. Please see the details in our article (Nomaguchi *et al.* 2008). It is not practicable to strictly quantify their values because they depend on the perception of a design process manager or a designer unlike physical parameters. This research is based on the premise that the accuracy of the quantification of the indistinct factors can be enhanced through interaction between the prediction by the simulation and the perceived actual results, even if the factors contain vagueness.

We thankfully acknowledge all OFRAC members for their contribution to the case analysis and study.

6.8 References

Cho SH, Eppinger SD (2005) A simulation-based process model for managing complex design projects. IEEE Transactions on Engineering Management, 52(3): 316-328

Clarkson PJ, Hamilton JR (2000) 'Signposting', a parameter-driven task-based model of the design process. In: Research in Engineering Design 12(1): 18-38

Eppinger SD, Whitney DE, Smith RP, Gebala D (1994) A model-based method for organizing tasks in product development. Research in Engineering Design 6(1): 1-13

Marca DA, McGowan CL (1988) IDEF0/SADT Business process and enterprise modeling, Eclectic Solutions Inc, San Diego, CA, US

Moder JJ, Phillips CR, Davis EW (1983) Project management with CPM, PERT & precedence diagramming. Van Nostrand Reinhold, New York, NY, US

Nomaguchi Y, Tsutsumi D, Fujita K (2008) Design achievement model for planning creative and concurrent design process. In: Proceedings of the ASME International Design Engineering Technical Conferences (IDETC/CIE2008), New York, NY, US

OFRAC Osaka University Formula Racing Club. Available at: http://ofrac.net/ (Accessed on 21 July 2009)

Ostergaard KJ, Summers JD (2007) Resistance based modeling of collaborative design, concurrent engineering. Research and Applications 15(1): 21-32

PMI Standards Committee (2004) A guide to the project management body of knowledge (PMBOK Guide). Project Management Institute, PA, US

Reich Y, Paz A (2008) Managing product quality, risk, and resources through resource quality function deployment. Journal of Engineering Design 19(3) 249-267

Chapter 7

The Effect of Uncertainty on Span Time and Effort within a Complex Design Process

S. Suss, K. Grebici and V. Thomson

7.1 Introduction

In large organisations that design complex products, it is the coordination of the creation and communication of information in the face of uncertainty that has become important in the competitive performance of the product development process. Improving management of the interdependencies between design tasks with an understanding of the various aspects of uncertainty offers an opportunity to substantially reduce span time and effort.

Product design is typically decomposed into a hierarchy of subtasks. This is done to cope with the complexity of the design task and to gain the benefits of specialisation. Many subtasks have dependencies between them which require management or coordination. These dependencies are primarily information that is generated within some subtasks and required by others; therefore, the performance of the entire design task is dependent on the progress of individual subtasks and the exchange of information between them. If the coordination of information is carried out efficiently and effectively, design projects have better results in terms of span time and effort. In order to find how to best coordinate the flow of information as the subtasks are being carried out, the structure and synchronisation of the information must be analysed.

However, this is difficult when the design process is highly iterative. A design task is likely to involve different amounts of work and represent different states, and so, design processes are iterated to achieve a succession of improvements, evolutions, or refinements on the way toward the final outcome. Additionally, the effective time spent for the completion of a task is often affected by the inherent uncertainty associated with the design information as it evolves during the process.

Hence, the authors tackled the following questions.

- How should levels of uncertainty be represented?
- How should the progress of work be modelled so that it bears the effect of uncertainty and the iterative nature of design?

- Is there any pattern as to how uncertainty affects the progress of work and overall project span time and effort?

To answer these questions, we developed a model of the development process of a complex product. The model considers several interdependent tasks performed in phases with a design review and the possibility of design version rework at the end of each phase. Tasks can be performed one after the other with partial or full overlap. We model the coordination of tasks with the use of policies governing information exchange between tasks, the attention of participants to communication, and the management of resources. The model evaluates the impact of various types of uncertainty associated with design that affect information flows, iterations, and thus, the progress of the design process. Executing the model with discrete event simulation yields results that point to strategies that can make new product development more robust, and the progress of the process more efficient in the face of uncertainty.

7.2 Characteristics of the Design Process

A design process is "an organized group of related tasks that work together to create a result of value" (Hammer, 2001) or a network of customer-supplier relationships and commitments that drive activities to produce results of value. An organised group of tasks can only produce results of value if coordination between the work being done in each task takes place. The coordination is reliant on an exchange of information (Galbraith, 1977). The information required by and exchanged between design tasks can take different forms. However, in this paper we consider that information flow consists of the following types of information: design information which can be specified directly, *e.g.* materials and geometry, performance information which is a consequence of design information, *e.g.* fatigue life and weight, and requirements which may constrain design or performance information, such as the requirements that the geometry of a component has to fulfil to meet the overall performance of the whole product. This follows the approach of others that have worked on analysis of information exchange in design processes (Krishnan, 1997; O'Donovan *et al.*, 2003; Bhuiyan *et al.*, 2004).

7.2.1 Process Structural Complexity

7.2.1.1 Task Interdependencies

The design process can be viewed as a system of interrelated activities, which are performed to increase knowledge or to reduce uncertainty about the design solution (Grebici *et al.*, 2008). During design, a system is decomposed into sub-systems, components and the parameters which define them.

Design processes are organised into concurrent streams of work which create product information. They are complex and difficult to understand in their own right, and this complexity is exacerbated by the interdependencies within and between the main domains: tasks, people and product. In practice, a perfect one-to-one mapping between the elements within and between the domains rarely exists in dynamic engineering design environments. Scarce or shared resources, multi-tasking, outsourcing, and dynamic or uncertain development demands, all make the many-to-one or a many-to-many mapping from one domain to another yield a model of potential interactions, not simply expected ones.

7.2.1.2 Process as Many Alternative Routes

Design processes are commonly considered as a complex net of tasks, information transfers and decision points. If all the design routes and decisions are made explicit, a complex network with important design alternatives should be considered (Clarkson *et al.*, 2000). In this paper, we do not attempt to model the decision points and task alternatives in a design process, but consider the process as a fixed group of interdependent tasks exchanging information.

7.2.2 Iterative Character of the Design Process

One of the most prominent features of a design activity is its iterative character. It is portrayed prominently in nearly every model of the design process (AitSahlia *et al.*, 1995; Krishnan, 1997). Motivation for studying the iterative behaviour of design tasks is not hard to find. The majority of development time and cost has been shown to be iterative activity (Cooper, 1993; Osborne, 1993). Therefore, methods to support more detailed task planning and control in complex design processes call for the construction of process models that capture design iteration among other process attributes (resource usage, information flow and milestones) (Wynn, 2007).

A design task is likely to involve different amounts of work and represent different states. Iterations in design are recognised to be of different natures: *repetition, incremental, progressive and feedback* (Safoutin, 2003). Progressive iterations are associated with the improvement or evolution of a design (Eppinger *et al.*, 1997; Safoutin, 2003). They correspond to processes that achieve a succession of improvements or refinements on the way toward the final outcome. The intermediate outcomes represent estimation of the final result.

In this paper we consider the possibility of iterations that occur when a task is being performed where not all of the required input information is available at the outset. The work begins with partial information which is uncertain and is updated periodically as the process unfolds. If two tasks are reciprocally dependent, then, the information received by the first task is used to generate information required by the second task. Rework occurs when a task has gone too far with the partial or preliminary information it has received, and upon receiving further information must rework some of what it has already done. This type of iteration is classified as progressive or incremental. We also consider similar iterations internal to the task itself.

Iteration that comes about as a result of a design review is also taken into account. Here, we reason that as tasks are completed and results reviewed, there is a likelihood that a task performed during a phase must be reworked in its entirety. This likelihood of design version rework diminishes with the completeness of information exchange that has taken place among the interdependent tasks in the process. This type of iteration brought about by design version rework is classified as feedback or repetition.

7.2.3 Uncertainty

There are several definitions of the term, uncertainty. This is mainly due to the fact that different types, causes and sources of uncertainty co-exist (Thunnissen, 2004). Perhaps one of the main distinctions regarding uncertainty in engineering design is between the lack of knowledge (epistemic) and random (aleatory) uncertainty (Oberkampf *et al.*, 2004; Mourelatos and Zhou, 2005). Epistemic uncertainty is derived from ignorance or incomplete information. Some epistemic uncertainty is reducible, for instance, via further studies, measurements or expert consultation. Random uncertainty describes the inherent variation associated with a system or environment such as dimensional variation in a production system or the variation in design task duration. This aleatory uncertainty cannot be reduced through more information or knowledge acquisition during the design process. Although gaining knowledge about variability may allow its influence to be mitigated through the design of systems that are adaptable, robust, flexible, *etc.* (Chalupnik *et al.*, 2009). In the engineering design process, the influence of variability may be managed through frequent information exchange, and one of the purposes of our analysis is to determine this.

Drawing on the literature of an earlier paper (Grebici *et al.*, 2008), two components of uncertainty within the design process were identified for inclusion in the model. Each of these aspects influences the iterative behaviour of design tasks.

- **Imprecision:** Aughenbaugh and Paredis (2006) define precision as the gap between certainty and a decision-maker's present state of information, i.e., the information currently available for decision-making. As design decisions are made, imprecision tends to progressively reduce until a final, precise value is determined (Antonsson *et al.*, 1995). For instance, greater precision justifies application of more sophisticated design tools and methods, which may require progressively greater effort (Evans, 1959).
- **Instability:** More unstable descriptions are more likely to change and instability may be increased by events which increase the likelihood that rework will be required. This component of uncertainty accounts for the gap between certainty and a state of precise information. For instance, the receipt of a change request can increase instability in design information which may require knock-off change (Eckert *et al.*, 2004).

The next sections present the arguments for the choice of instability and imprecision as components of uncertainty in the proposed model.

7.3 Modelling Process Progress

The term 'process progress' is similar to the term 'work state' (Carrascosa *et al.*, 1998); these terms refer to the effective time spent for the completion of a task. Wood *et al.* (1990) propose that as a design progresses the level of design imprecision is reduced, although a degree of stochastic uncertainty usually remains. Having a high level of confidence in design information means that the information is detailed, accurate, robust, well-understood, and physically realistic (O'Donovan *et al.*, 2003). Also, progress in a design process refers to a process that achieves a succession of improvements or refinements on the way toward the final outcome (Safoutin, 2003).

Hykin and Laming (1975) refer to work progress as a process in which the level of uncertainty in the artefact is reduced as the design progresses. We incorporated this reasoning in our model in the way in which we calculate the variation in uncertainty in precision with the state of progress in a task. According to Carrascosa *et al.* (1998) the S-shape is the best shape function to represent task completion as it can embody a growth phenomenon similar to learning.

In this paper we use a Gompertz function to model the reduction in uncertainty as a function of work progress because of its flexibility in allowing for different rates of approach to the lower and upper asymptotes at activity start and end. The work progress curve can be transformed into an uncertainty reduction curve by subtracting the Gompertz function from its upper asymptote (Equation 7.1).

7.4 Description of the Model

A task based approach for process modelling is used in this work since it allows the link between information flow characteristics, functions, and objectives (*e.g.* the fulfilment of product performance parameters).

For a particular project (product to be developed and organisational capability), we postulate that there is a certain total amount of information that must be processed. We model the information flows by characterising the information dependency between each pair of tasks i and j as made up of the initial uncertainty in information input to task j from task i and by the initial sensitivity of task j to changes in this information input (Krishnan, 1997; Yassine *et al.*, 1999). Following the work of the latter authors, we use the product of these two characteristics to arrive at a value for the dependency strength between tasks.

Thus, we define a project's information dependencies using a dependency structure matrix (DSM) $\boldsymbol{D}$ whose elements are defined as follows:

$$D_{ij} = U_{ij} * S_{ij} \tag{7.1}$$

where U_{ij} is the initial uncertainty of the information input from task i in development team j, and S_{ij} is the initial sensitivity of the output of task j to changes in information from task i. Using values of 0, 1, 2, and 3 to describe levels of uncertainty and sensitivity of none, low, medium and high for U_{ij} and S_{ij}, the elements D_{ij} can take on values of 0 through 9.

We further hypothesise that the total amount of information that must be communicated between interdependent tasks is directly proportional to their dependency strength D_{ij}. This can be understood by considering that, if there is greater uncertainty in an activity's output, it is likely that more estimates of the output information will need to be generated and communicated to downstream activities before the design activity is completed. Similarly, if there is greater sensitivity to another activity's output, it is likely that more information will need to be transferred before the linked activities arrive at a jointly satisfactory solution. For each increased level of uncertainty, the effect of sensitivity is magnified and vice versa; so, we assume that the effects of these characteristics are multiplicative as indicated in Equation 1. Empirical results supporting this hypothesis were published by Allen (2007).

Design tasks are modelled through the execution of technical work, the development of information, and the communication of information that occurs in each task. Thus, we model a task as composed of a work subtask and communication subtasks.

7.4.1 The Work Subtask

Technical work requires the development team (resource) to perform the work subtask for a specific amount of time chosen from a triangular distribution with given minimum, median, and maximum limits. As the technical work progresses, information is developed with a value of uncertainty related to the state of the task producing it. The state, defined as a linear combination of the technical work fraction and the input information fraction, is related to the precision aspect of uncertainty as previously discussed and is modelled with the following equation:

$$\varepsilon_{\tau i} = 1 - \mathbf{exp}(-b_i * \mathbf{exp}(-c_i * t_{\tau i})) \tag{7.2}$$

where $\varepsilon_{\tau i}$ is the precision part of uncertainty divided by the initial value of uncertainty for task *i,* and $t_{\tau i}$ is the state of task *i* at time *T* during the simulation. Equation (7.2) is the Gompertz function subtracted from its upper asymptote (here equal to one) and where b_i, c_i are input coefficients that define the shape of the curve. This function has some of the characteristics of an S-shape with the steepness of reduction of uncertainty controlled by the choice of the coefficients b_i, c_i for each task *i*. The state of the ith task $t_{\tau i}$ at time *T* is defined as:

$$t_{\tau i} = ½ * (W_{\tau i}/W_{D i} + I_{\tau i}/\Upsilon_i) \tag{7.3}$$

$W_{\tau i}$ is the amount of technical work done until time *T* in the simulation by task *i*; $W_{D i}$ is the total amount of technical work required to be done by this task (initially determined by sampling a triangular distribution at the onset of the task); $I_{\tau i}$ is the amount of input information received from all other activities until time *T*; and Υ_i is the total amount of input information required to be received by task *i* from all other activities. This total amount of input information is determined for a given scenario of initial uncertainty of information output and sensitivity to the information input of the process activities, and is further elaborated in (Suss and Thomson, 2009).

The instability part of uncertainty is modelled as follows:

$$\varphi_{\tau i} = \varepsilon_{\tau i} * M_i * \mathbf{UNIF}(0,1) \tag{7.4}$$

where $\varphi_{\tau i}$ is the ratio of the instability part of uncertainty at time *T* to the initial uncertainty of task *i*; M_i is a scaling factor; and **UNIF**(0,1) is a sample chosen from the uniform probability distribution between 0 and 1. Since the total uncertainty is the sum of $\varphi_{\tau i}$ and $\varepsilon_{\tau i}$ the value of M_i represents the initial magnitude of the uncertainty of task *i*.

7.4.2 The Communication Subtasks

As information is developed it must be prepared for communication and this requires time and effort by the resource. The time required for a unit resource to prepare a unit of information for communication is chosen from a triangular distribution with parameters representing the minimum, median, and maximum values needed. Similarly, a development team receiving input information requires time to read and interpret it. This 'read' processing time is modelled in a similar way to the 'prepare' time. The work, prepare and read subtasks within a task are all performed by the same development team which is modelled as a finite resource.

Since the work, prepare and read subtasks occur in tandem, the ith work subtask is broken down into discrete time periods ΔT_i. After each ΔT_i, the ith team checks if there are any information units in their 'inbox' to process. If there are more than a minimum number, QM_i, the waiting communication work is performed prior to returning to the work subtask where it left off. The interval ΔT_i and QM_i are determined by the project coordination policy that is being studied. For example, if we wish to model a development team that routinely pauses in its technical work and checks if there are communication subtasks to perform once a week, the input value of ΔT for this task is 40 hours. Furthermore, if we model the team to only perform the communication subtasks when there are more than 5 items waiting to be processed, the input value of QM for this team is 5. In a project where there are critical requirements for timely information exchange between several tasks, we may prefer to model the team's behaviour with values of ΔT and QM of 8 and 1 respectively.

Information developed within a task as it progresses from state to state in the discrete event simulation is communicated to other development teams performing activities that are dependent on this information. Assuming that a certain amount of information must be sent from task i to other tasks, this information is broken into discrete units or entities. These entities are created by task i according to the state of progress in its work. An information entity is created each time the work progress fraction (work done divided by work required to be done) has reached or exceeded the fraction required, i.e., 10% of the information entities are created in the task when 10% of the work is done.

These information entities are held by the task and sent to other activities only when a group of them has been created. The communication frequency of task i is modelled by setting the number of entities in a group with an input parameter, called MPG_i. Thus, if MPG_i is set to 1, an information entity is immediately prepared for communication after it has been created. This corresponds to the highest frequency of communication. If MPG_i is 50, there must be 50 entities in the group before they are prepared for communication. If 50 is the total amount of information entities that are required to be sent from activity i to all other activities that are dependent on it, then, all information is in effect held until the activity has almost completed the task before being sent out. If two tasks with reciprocal dependencies are modelled in this way, the simulation corresponds to the 'throwing results over the wall' scenario that is sometimes observed.

7.4.3 Modelling Iteration

Values of $\varphi_{\tau\,i}$ and $\varepsilon_{\tau\,i}$ are attributes of each information entity. When information is 'read' by the receiving activity *j,* these attributes are used to calculate the average total input uncertainty $\bar{U}^k_j$ and the average input value of $\acute{\varepsilon}^k_j$ prior to the k_{th} work time period in activity *j*:

$$\bar{U}^k_j = \sum_i (\varphi_{\tau i}{}^k + \varepsilon_{\tau i}{}^k) \,/\, N_k \qquad \text{sum from } i=1 \text{ to } N_k \tag{7.5}$$

$$\acute{\varepsilon}^k_j = \sum_i (\varepsilon_{\tau i}^{\,k}) \,/\, N_k \qquad \text{sum from } i=1 \text{ to } N_k \tag{7.6}$$

where N_k is the number of information entities arriving at task *j* prior to work cycle *k*.

Prior to commencing a cycle of work, $\bar{U}^k_j$ is compared to $\bar{U}^{k-1}_j$. We reason that if the new value of input uncertainty is higher than the previous one, rework must be done. This is because the previous cycle of work was based on a perceived value of precision in input information that was overly high compared to the new one, and thus, led to more work or the wrong work being done than was warranted by the level of uncertainty that subsequently became apparent.

The amount of incremental rework that is required in task *j* is calculated as follows:

$$W_{Dj}{}^{new} = W_{Dj}{}^{old} + W_{\tau j}{}^{k} - W_{\tau j}{}^{m} \tag{7.7}$$

where $W_{\tau j}{}^{m}$ is the amount of work done in subtask *j* until the beginning of the work cycle *m* where the value of $\acute{\varepsilon}^m_j$ is greater than $\bar{U}^k_j$. Thus, the amount of rework that is generated corresponds to the amount of work done by this task since the perceived imprecision of input information is greater than the new total input uncertainty.

The simulation unfolds as follows: each task starts to perform its work subtask according to the sequence set in the modelled process (full, partial or no overlap); as the work progresses, discrete information units are created with uncertainty based on the state of the task at each discrete time step; information units are sent to other activities according to the frequency of information exchange being modelled; as work is progressing, if new information is not received in time, the task's work subtask stops and waits for more input information. The criterion for this is simply:

$$W_{\tau j} > W_{Dj} * I_{\tau j} / \Upsilon_j \tag{7.8}$$

Once information is received the work subtask continues, but additional work may be required according to the scheme described in Equation 7.7.

The activities form part of a phase in the process with a design review modelled at the end of a phase. The fraction of completeness of the work subtask and information exchange are used to determine the likelihood of the project to continue to the next phase or to perform design version rework. If design version rework is required, a learning factor reduces the work and information exchange requirements for each rework cycle. More details of these aspects of the model are in (Suss and Thomson, 2009).

7.5 Results and Analysis

Analysis of simulation results provides insights about the effects of epistemic uncertainty reduction, random uncertainty, frequency of information exchange, resource allocation, and degree of task overlap on project effort and span time. Data on queue lengths and waiting times, rework generated due to churn and due to design versions provides further understanding of how the complex interplay of the properties of these scenarios results in the global outputs of span time and effort.

As described in Sections 7.3 and 7.4, a Gompertz function was used to determine the epistemic uncertainty ε at every state of each task. Examples of two Gompertz functions of uncertainty reduction are shown in Figure 7.1 where the coefficient b has a constant value of 5, and c is set at 4 and 20. The higher the value of c, the greater the steepness of the Gompertz curve. Indicated on these charts is the contribution of random uncertainty φ to the total uncertainty with a scaling factor M of 0.1.

Simplified scenarios of 3 tasks sequentially and reciprocally dependent in a two phase development project were simulated. Sufficient replications were made of each simulation to ensure 95% confidence levels. We carried out these simulations to validate the model for scenarios that have been addressed by other researchers and that have been observed in practice. Scenarios with more tasks and more complex sequencing and dependency relations could be simulated to investigate processes of greater complexity.

In Figure 7.2, results are shown for overall effort in a two phase project versus overall span time for a process of three tasks with sequential dependence (task 3 requires information from tasks 1 and 2, task 2 requires information from task 1 and task 1 does not require any information from other tasks). The bottom curve had the rapidly reducing epistemic uncertainty function of figure 1 with $M = 0, 0.05, 0.1$ (These are indistinguishable), and the other three curves used the more slowly reducing epistemic uncertainty function in that figure with $M = 0, 0.05, 0.1$ (lowest to highest).

The results indicate a reduction in span time as overlapping is increased for all cases. When the rate of epistemic uncertainty reduction is rapid, there is a 33% reduction in span time with no increase in effort when tasks are partially overlapped, and there is a 50% reduction in span time with a 5% increase in effort when the tasks are fully overlapped. However, when the epistemic uncertainty profile diminishes more slowly (3 upper curves in figure 2), the penalty in effort is high when tasks are partially overlapped with a small reduction in span time. In the fully overlapped case, the penalty in effort rises rapidly and the reduction in span time diminishes depending on the value of the instability scaling factor.

In Figure 7.3, scenarios of three highly interdependent tasks (each task required input information from each of the other tasks) were simulated. The tasks were fully overlapped. Each curve in figure 3 shows results generated for cases of increasing frequency of information exchange with different profiles of epistemic uncertainty reduction, but with constant values of the instability scaling factor (M=0.05). The results indicate that when information is exchanged only once near the end of the task, span time and effort are almost the same regardless of the uncertainty profile. However, when information is exchanged more often, there is a significant reduction in span time with an increase in effort. This increase in effort is more rapid when the rate of epistemic uncertainty reduction is slow. Where the rate of uncertainty reduction is rapid, there is an actual reduction in effort. Interestingly, there is a point where higher frequency of communication only serves to increase the effort in the project without any gain in span time.

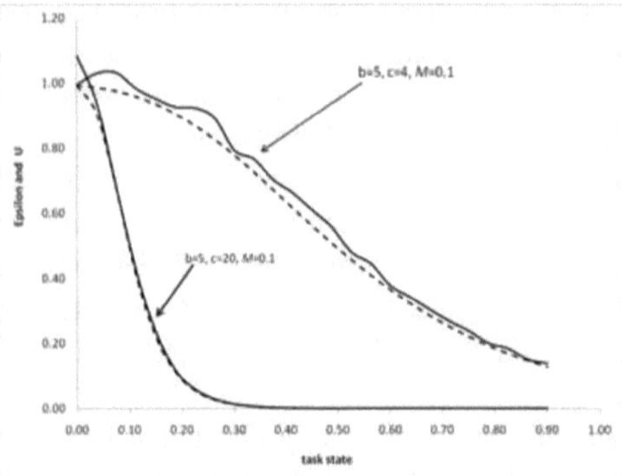

Figure 7.1. Examples of uncertainty reduction functions used in the simulation model

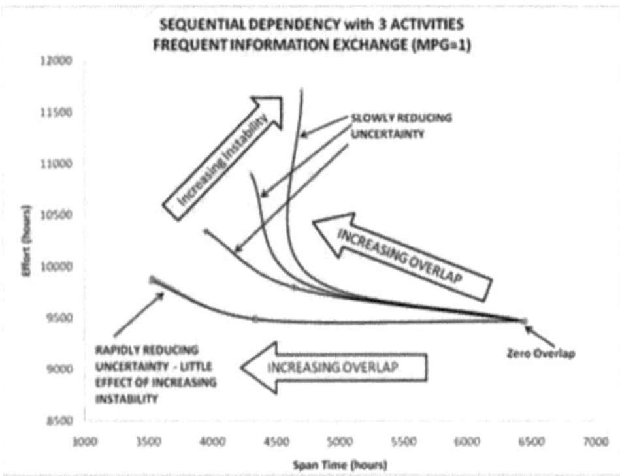

Figure 7.2. Effort versus span time with increasing overlap of tasks for several cases of uncertainty reduction profiles and instability with sequentially dependent tasks

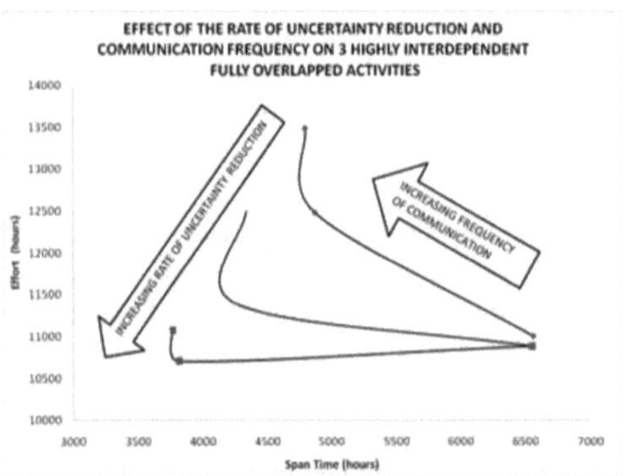

Figure 7.3. Effort versus span time with increasing frequency of communication for several cases of uncertainty reduction profiles with highly interdependent tasks

7.6 Conclusion

In this paper we describe a stochastic, task based, executable model of a complex product development process. Simulation results are shown capable of providing relevant insights with regard to the effect of uncertainty on process performance as well as with regard to improved coordination policies (information exchange resource allocation, and/or task overlapping). Simulations of interdependent design processes indicate behaviour in line with that observed in practice and other research (Bhuiyan, 2004; Eckert *et al.*, 2004; Krishnan, 1997; Terwiesch, 2002).

The model relates epistemic uncertainty to the state of progress in a design task which in turn is related to the ongoing work and information exchange that takes place as the simulation unfolds. Incremental and feedback types of iterations are emergent in the simulation when aleatory uncertainty and delays due to resource constraints, queuing effects, and communication protocols influence the information exchange between interdependent activities, and thus, the progress and completeness of the activities.

More complex scenarios of product development projects with many more tasks can be analysed with this model providing insights into the drivers of project cost and span time as a result of coordination practices. The approach adopted here that follows the information flows and relates them to the progress in each task can deal with the complexity of real design processes since the fundamental 'rules' implemented in this model still hold.

Future research should investigate the reduction of epistemic uncertainty at every state of each task while taking into account the content of the design tasks themselves. For instance, during early design phases, more progression types of iteration are used as the designers progressively define the geometry of the main components of the product (without detailing them), whereas in detail design more incremental iterations are applied as the designers improve the detailed geometry of a component by applying the same activities to obtain better estimates of the outcome. The difference between progression and incremental iterations should be reflected in the epistemic uncertainty reduction curve by considering larger and few repetitions, and hence, slower epistemic uncertainty reduction in the case of progression iterations. Furthermore, although imprecision and instability seem to be relevant uncertainty components that can affect the pattern of progress, and thus, process behaviour, other components such as indecision and inaccuracy should be studied (Grebici et al., 2008).

7.7 References

AitSahlia F, Johnson E, Will P (1995) Is concurrent engineering always a sensible proposition IEEE Transaction on Engineering Management 42(2): 166-170

Allen TJ (2007) Architecture and communication among product development engineers. California Management Review 49(2): 23-41

Antonsson E, Otto KN (1995) Imprecision in engineering design. ASME Journal of Mechanical Design 117(B): 25-32

Aughenbaugh JM, Paredis CCJJ (2006) The Value of Using Imprecise Probabilities in Engineering Design. Transactions of the American Society of Mechanical Engineers Journal of Mechanical Design 128(4): 969-980

Bhuiyan N, Gerwin D, Thomson V (2004) Simulation of NPD process and performance. Management Science 50(12) 1690-1703

Carrascosa M, Eppinger SD, Whitney DE (1998) Using the design structure matrix to estimate product development time. In: Proceedings of the ASME International Design Engineering Technical Conference (IDETC/CIE1998), Atlanta, GA, US

Chalupnik MJ, Wynn DC, Clarkson PJ (2009) Approaches to mitigate the impact of uncertainty in development processes. In: Proceedings of the 17th International Conference on Engineering Design (ICED'07), Paris, France

Clarkson PJ, Melo AF, Eckert CM (2000) Visualisation of routes in design process planning. In: Proceedings of the International Conference on Information Visualisation (IV2000), London, UK, pp 155-164

Cooper KG (1993) The rework cycle: benchmarks for the project manager. Project Management Journal 24(1):17-21

Eckert CM, Clarkson PJ, Zanker W (2004) Change and customisation in complex engineering domains. Research in Engineering Design 15(1): 1-21

Eppinger SD, Nukala MV, Whitney DE (1997) Generalised models of design iteration using signal flow graphs. Research in Engineering Design 9: 112-123

Evans JH (1959) Basic design concepts. American Society of Naval Engineers Journal 71(4): 671-678

Galbraith J (1977) Organization Design. Addison Wesley Publishing Company, Reading, MA, US

Grebici K, Wynn D, Clarkson JP (2008) Modelling the relationship between the uncertainty levels in design descriptions and design process duration. In: In: Proceedings of the International Conference on Integrated Design and Manufacturing in Mechanical Engineering (IDMME 2008), Beijing, China

Hammer M (2001) Seven insights about processes. In: Proceedings of the Conference on Strategic Power Process Ensuring Survival Creating Competitive Advantage, Boston, MA, US

Hykin DHW, Laming LC (1975) Design case histories: Report of a field study of design in the United Kingdom Engineering Industry. In: Proceedings of the Institute of Mechanical Engineers 189(23): 203-211

Krishnan V (1997) Managing the Simultaneous Execution of Coupled Phases in Concurrent Product Development. IEEE Transactions on Engineering Management 43(2): 210-217

Mourelatos Z, Zhou J (2005) Reliability estimation and design with insufficient data based on possibility theory. American Institute of Aeronautics and Astronautics Journal 43(8): 1696-1705

Oberkampf WL, Helton JC, Joslyn CA, Wojtkiewicz SF, Ferson S (2004) Challenge problems: uncertainty in system response given uncertain parameters. Reliability Engineering and System Safety 85(1-3): 11-19

O'Donovan BD, Clarkson PJ, Eckert CM (2003) Signposting: Modelling uncertainty in design processes. In: Proceedings of the 14th International Conference on Engineering Design (ICED'03), Stockholm, Sweden

Osborne SM (1993) Product development cycle time characterization through modeling of process iteration, MSc Thesis, Massachusetts Institute of Technology, Boston, MA, US

Safoutin MJ (2003) A methodology for empirical measurement of iteration in engineering design processes. PhD Thesis, University of Washington, Seattle, WA, US

Suss S, Thomson V (2009) A model of a complex new product development process. In: Proceedings of the International Conference on Industrial Engineering and Systems Management (IESM'09, Montreal, Canada

Terwiesch C (2002) Exchange preliminary information in concurrent engineering: Alternative coordination strategies. Organization Science 4: 402-419

Thunnissen DP (2004) Propagating and mitigating uncertainty in the design of complex multidisciplinary systems. PhD Thesis, California Institute of Technology, Pasadena, CA, US

Wood KL, Antonsson EK, Beck JL (1990) Representing imprecision in engineering design: comparing fuzzy and probability calculus. Research in Engineering Design 1(3/4): 187-203

Wynn DC (2007) Model-based approaches to support process improvement in complex product development. PhD thesis, Cambridge Engineering Design Centre, University of Cambridge, Cambridge, UK

Yassine AA, Falkenburg D, Chelst K (1999) Engineering design management: an information structure approach. International Journal of Production Research 37(13): 2957-2975

Chapter 8

Evaluating the Positive and Negative Impact of Iteration in Engineering Processes

H.N. Le, D.C. Wynn and P.J. Clarkson

8.1 Introduction

Iteration is often considered to be one of the most prominent and fundamental characteristics of engineering design processes. For instance, the high-level design process models proposed by Pahl and Beitz (1996), French (1998) and Pugh (1996) symbolise iteration through feedback arrows occurring between different design process stages. Osborne (1993) conducted a study in the semiconductor industry to measure the magnitude of iteration in product development; his findings suggest that a significant share of development time and cost can be allocated to iterative activities (Browning, 1998). Many other authors have highlighted the central role of iteration in the design process, and hence indicated its importance in product development. Consequently, both industry and academia have a strong interest in better understanding the nature of iteration, including its causes and effects on design processes, so it can be effectively managed to help ensure development time and cost is as low as possible.

This paper draws on the literature of engineering design processes to synthesise a new explanatory framework for understanding the causes and effects of iteration. This framework forms the basis for a quantitative analytic approach which can be used to better understand, and hence better manage the positive and negative influences of iteration on process performance. In particular, our research is motivated by the need to:

- understand the different causes and purposes of design iteration;
- understand the effects of different iteration forms on process performance;
- thereby give managers the ability to explore ways to reduce the negative impact and enhance the positive impact of iteration in their processes, through manipulation of process management levers and through process engineering.

The argument is structured as follows. We begin by discussing the different causes and effects of iteration. Then, after reviewing existing frameworks which support analysis of iteration, we synthesise a new explanatory framework which considers all the main issues we identified in the literature and presents them in a way helpful to explore the challenges outlined above. We discuss some approaches which have been proposed to mitigate the negative effects of iteration and enhance the positive effects, highlighting the need for a more integrated approach to fully explore the impact of the interrelated levers which can be used to manage iterative behaviour and ultimately process performance. We then outline an analytical approach we are developing to resolve this shortcoming.

8.2 Causes and Effects of Iteration

The causes of iteration are both external and internal to the design process. External causes of iteration arise from technology and market uncertainties (Eisenhardt and Tabrizi, 1995). For instance, many new products have to include novel technologies whose performance and viability risks are not fully determined when early design takes place. Therefore, an iterative process of exploration is required where alternative designs are developed and considered in parallel and much testing is conducted before a concept can be selected. Even when the design has proceeded further, companies may have to make changes and adapt new technologies to their design in order to stay competitive in a changing environment (Fricke and Schultz, 2000). It is also common that customers modify their requirements during the design process after an interim solution is communicated (Costa, 2004). With each change in requirements definition or technology evaluation, designers must revisit some work, *i.e.* iterate.

Internal to the design process, there are also many events which can cause iteration to occur during design. The design of large-scale products is equivalent to solving a set of complex and coupled problems (Clarkson *et al.*, 2000) with conflicting objectives (Denker *et al.*, 2001) where the solution-finding process includes many cycles of exploration and convergence. On a detailed level, the interdependencies between components result in interdependent tasks that influence each other's input; these interdependencies can thus require tasks to be revisited to achieve a technical solution (Smith and Eppinger, 1995). Similarly, iteration is likely to occur when interface or system integration issues are considered (Hoedemaker *et al.*, 1999), because new information may surface or it may become clear that design objectives are not met when testing is performed. Additionally, due to the complex nature of engineering design problems, errors and mistakes cannot be entirely avoided. Time usually elapses before such errors are detected and corrected by reworking the affected design, potentially involving many intermediate activities in iteration too (Cho and Eppinger, 2005).

In addition to these technical issues, the management of design processes can also be a process-internal cause of iteration. Under the pressure to reduce development lead time, managers often choose to schedule dependent tasks concurrently based on early release of information (Joglekar *et al.*, 2001). As a consequence, tasks need to be repeated when input information is finalised and the amount of rework may vary according to the chosen degree of concurrency (Krishnan *et al.*, 1997; Roemer *et al.*, 2000). Rework is also likely to occur when deadlines are set in a way that available time is not adequate to workloads, causing high work intensity and overtime for designers, and resulting in higher chances of flawed work (Lyneis and Ford, 2007). Another human related aspect is the nature of concealing problems and need for rework under schedule pressure, causing even more rework to be required later in the design process (Ford and Sterman, 2003).

Much of the product development literature thus concludes that iteration can significantly affect development time in a negative way. Several authors have observed iteration as a key driver of schedule slippage in product development, especially in large-scale projects (Adler *et al.*, 1995; Smith and Morrow, 1999). In his fieldwork, Osborne (1993) found that 13% to 70% of total development time for semiconductor development could be considered as iteration (Browning, 1998). Other authors have demonstrated, through simulation, how undiscovered rework creates a falsely optimistic perception of progress and that later discovery of this rework can lead to more effort being required to fix flawed tasks (Cooper and Kleinschmidt, 1996; Ford and Sterman, 2003). In addition to the primary effects, rework can also create secondary effects - such as increased work intensity and schedule pressure - that influence productivity and work quality in such a way that the overall progress can be affected as well (Lyneis and Ford, 2007). Similarly, Terwiesch *et al.* (2002) conclude that iteration also tends to increase development cost, as engineering hours increase.

Although common, not all authors present iteration and its effects on process performance in a negative way. For instance, results from a survey by Eisenhardt and Tabrizi (1995) indicate that development processes with "more design iterations" tend to require less time. Fricke and Schultz (2000) argue that the quality of the product can be improved by allowing late technology changes to be incorporated through iteration, and by doing so to stay competitive.

In summary, the product development literature suggests that iteration has many different causes - both internal and external to the process, and arising from both technical and management issues. Authors focusing on the negative effects of iteration refer to unproductive rework, which can be caused by factors such as flawed design and insufficient quality assurance. Authors reporting the positive effects focus on iteration as being necessary to systematically explore and understand the complexity of design problems and their potential solution space, leading to a more efficient solution-finding progress.

8.3 A framework of Iteration Causes and Effects

The apparent complexity of iteration indicates that exploiting opportunities to reduce the negative impacts of iteration and enhance the positive effects requires a comprehensive understanding of iteration and its impact on process behaviour. Several frameworks have been proposed in the literature to assist with this. For instance, a functional classification of iteration can be found in the differentiation between *diagnostic* iteration (*i.e.* defining and evaluating tasks) and *transformative* iteration (*i.e.* synthesising new information) (Adams *et al.*, 2003). Smith and Eppinger (1997) classify iteration between tasks as occurring in *parallel* or *sequential*, according to the structure of information (inter)dependencies between those tasks. Browning (1998) differentiates between *planned* iteration, which is intentionally executed to create useful information, and *unplanned* iteration, which is caused by emerging information at the wrong time in the process. Similarly, Costa and Sobek (2003) classify iteration according to its intended contribution to the design, with *design* and *behavioural* iteration providing positive progress to design whereas *rework* iteration only repeats work to correct flawed design. Wynn *et al.* (2007) highlight the complexity of iteration through discussing six "non-orthogonal perspectives" by which it can be viewed: *exploration, convergence, refinement, rework, negotiation* and *repetition*. They argue that iteration is difficult to classify since it is not an entirely objective phenomenon, but has different meanings and importance depending on domain- and stakeholder-specific lenses. For instance, whether iteration is considered to generate useful information or just to correct errors is likely to depend on the perspective of the person considering it; some new information is almost always generated when work is repeated and could influence subsequent work.

Each of these existing frameworks offers different insights into iterative behaviour in product development. In an attempt to synthesise these views and to support further study of the questions outlined in the introduction, the causes of iteration revealed through the literature were clustered according to their contexts within the design process (Table 8.1). Causes in the *Technology and market uncertainty* cluster are related to process-external influences; causes falling into the *Process management/human factors* cluster are related to the project environment; and causes in the *Design issues* cluster are related to design activity. Each cause was then matched to reasons for iteration and effects of iteration, as shown in the table. To summarise:

- **Iteration can stem from various sources** - As indicated in the framework, iteration can stem from process external influences, the project environment and design issues. The high number of possible causes complicates the cause-and-effect relationship;
- **Some of the causes can result directly from process management** - Unlike the causes in *Technology and market uncertainty* and *Design issues*, some causes in *Process management/human factors* are under management control and influence, and could, hence, be directly mitigated if the cause-and-effect relationship is well understood;
- **Effects of iteration can be complex and ambiguous** - The ambiguity of the effect of iteration is threefold: Firstly, iteration due to *technology/market uncertainty* and *process concurrency* has the potential to positively affect development lead time; secondly, some iteration may distort schedules; and thirdly, some iteration may reduce work quality and productivity. The effect of iteration is ambiguous in that it is not clear to what extent iteration affects the lead time, schedule or work quality and productivity, directly and indirectly. This is complicated by the interdependencies between causes of iteration which occur simultaneously.

Table 8.1. Summary of iteration causes and effects

	Causes of iteration	Reasons for iteration	Effects of iteration on process
Technology and market uncertainty	**Technology and Market uncertainty** *e.g.* Eisenhardt and Tabrizi, 1995; Loch *et al.*, 2001	- explore design space, build intuition and flexible options - increase robustness - avoid fixation	- potentially mitigate later rework possibilities, thus reduce development lead time - additional work requires additional time and resource
	Technology novelty *e.g.* Fricke and Schultz., 2000	- adaptation of new technology to stay competitive	- additional work requires additional time and resource - potentially distort schedule - potentially reduce work quality and productivity
Process management/ human factors	**Process concurrency** *e.g.* Krishnan *et al*, 1997; Joglekar *et al.*, 2001	- partial reattempt of task with final input for completion	- potentially reduce development lead time - additional work requires additional time and resource
	Concealing known problems *e.g.* Ford and Sterman, 2003	- solve earlier issues and relating backlog impact	- additional work requires additional time and resource - potentially distort schedule - potentially reduce work quality and productivity
	Work intensity, fatigue *e.g.* Lyneis and Ford, 2007	- modify flawed design	- additional work requires additional time and resource - potentially distort schedule - potentially reduce work quality and productivity
Design issues	**Complexity of design, multiple influencing factors** *e.g.* Denker *et al.*, 2001	- incremental incorporation of influencing factors into solution finding process - converge to a solution	- additional work requires additional time and resource
	Conflicting requirements *e.g.* Clarkson *et al.*, 2000	- refine parameter values to meet conflicting requirements	- additional work requires additional time and resource
	Problems in integrating system's components *e.g.* Hoedemaker *et al.*, 1999	- solve problems in system components' integration	- additional work requires additional time and resource - potentially distort schedule - potentially reduce work quality and productivity

To further explore the interdependencies between factors influencing iteration, the issues revealed through the literature were synthesised into a network of causes and effects. This is summarised in Figure 8.1, where each arrow represents the influence of a source factor on a sink factor. A negative sign indicates that an increase in occurrence of the source factor leads to a decrease in occurrence of the sink factor, and vice versa. A positive sign indicates the opposite relationship. The dashed, non-shaded factors are considered to be exogenous influences on iteration, *i.e.* they are not under management control and/or influence. The management policy-related and process architecture-related factors are considered to be partially under management control and influences.

In addition to the complex interdependencies highlighted by this diagram, the figure shows that the network of iteration causes and effects involves causal chain loops with reinforcing character. For instance, flawed design could lead to unplanned rework and, thus, additional work to be done. This, in turn, would cause the time pressure on designers to increase further, resulting in more overtime and fatigue. Under such conditions, the work quality and productivity tend to decline and, consequently, more errors may be made.

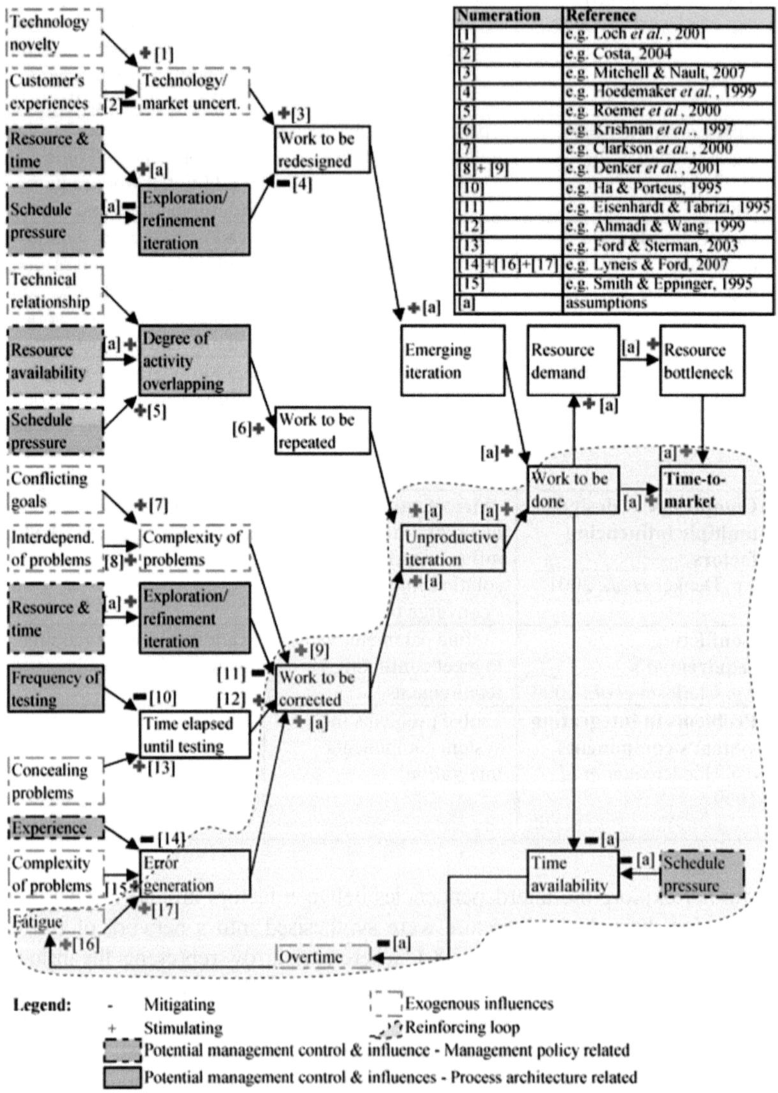

Numeration	Reference
[1]	e.g. Loch *et al.*, 2001
[2]	e.g. Costa, 2004
[3]	e.g. Mitchell & Nault, 2007
[4]	e.g. Hoedemaker *et al.*, 1999
[5]	e.g. Roemer *et al*, 2000
[6]	e.g. Krishnan *et al*., 1997
[7]	e.g. Clarkson *et al.*, 2000
[8]+[9]	e.g. Denker *et al.*, 2001
[10]	e.g. Ha & Porteus, 1995
[11]	e.g. Eisenhardt & Tabrizi, 1995
[12]	e.g. Ahmadi & Wang, 1999
[13]	e.g. Ford & Sterman, 2003
[14]+[16]+[17]	e.g. Lyneis & Ford, 2007
[15]	e.g. Smith & Eppinger, 1995
[a]	assumptions

Figure 8.1. Network of causes and effects of iteration

8.4 Managing Iteration

Appropriate management of the design process can mitigate the influence of some unproductive iteration and enhance the impact of positive iteration. In the product development literature, many approaches have been proposed which aim to support this through analyses of iteration's impact on a process' performance according to its architecture (*e.g.* the dependencies between activities, resources *etc.*). Some examples of these approaches are discussed below.

Approaches mitigating unproductive iteration arising from management issues. Companies often overlap development activities to compress total lead time. A number of authors discuss how this can result in more rework effort and how this can be mitigated. For instance, Krishnan *et al.* (1997) propose a framework where the evolution and sensitivity properties of the exchanged information between two activities are considered to determine the optimal degree of overlapping between these two activities with the objective to minimise rework effort. They address four basic cases of the evolution and sensitivity properties in their framework and derive an overlapping strategy for each case. Loch and Terwiesch (1998) model the influence of design uncertainty and dependency which make overlapping less attractive in their presence, *i.e.* the time saved through overlapping decreases. Roemer *et al.* (2000) map the trade-off between saving time and incurring cost through overlapping of activities and provide a plot of Pareto-optimal overlapping strategies to support decision making.

Approaches to mitigate unproductive iteration arising from design issues. Large-scale engineering design is highly complex and involves solving of coupled problems under conflicting goals and constraints. Thus, it requires many iteration cycles to converge to a solution. An appropriate approach to problem solving can help to reduce iteration effort and by doing so minimise "unnecessary" iteration. Many authors suggest application of a breadth-first approach to explore the possible design space at a higher design level where alternative solutions are considered, in order to achieve a more comprehensive evaluation of feasibility and a thorough understanding of design interface issues (*e.g.*, Hoedemaker *et al.*, 1999; Powell *et al.*, 1999; Sobek *et al.*, 1999). Testing also plays an essential role from the beginning and is recommended to be conducted earlier, for longer and more comprehensively to help discover errors in time to prevent rework. Eisenhardt and Tabrizi (1995) find support for the positive correlation between comprehensive testing and shorter product development time in their survey. On the operational level, Ha and Porteus (1995) apply an analytical model to estimate the optimal review period for different degrees of concurrent design. They argue that it is more beneficial to conduct frequent reviews if parallel development can be facilitated or the probability of design flaws is high. They also show that both infrequent and too frequent process review would result in negative impacts on the process development time. Similarly, Ahmadi and Wang (1999) come to the conclusion that an over-specified, as well as insufficient, stage confidence requirement level for process review in the early stage can cause extended lead time. In their model the Markov chain is applied to find the review path with the lowest cost while meeting the required level of confidence in the design.

Approaches to enhance the positive effect of iteration. In general, it is agreed that comprehensive exploration and refinement iteration at the early stage of the design process will offer "more chances for a hit" (Eisenhardt and Tabrizi, 1995) and help to cope with uncertainties and changing environment (Loch *et al.*, 2001). A thorough exploration of design alternatives can help to reduce the cost and effort of requirement change implementation at later stages in the design process. Exploring design alternatives also enhances the evaluation of each single one when design alternatives can be explored thoroughly (Costa, 2004; Gries, 2004). Various approaches such as TRIZ have been developed to support identification of design alternatives (Kim and Cochran, 2000). However, few studies have quantified the positive influence of productive iteration.

To summarise, many analytic approaches have been proposed to help understand, and therefore manage iteration in the literature. The strategies and policies derived from these approaches provide some guidance on how to cope with iteration and its effects on the process. However, most focus on a relatively small number of the iteration perspectives and influencing factors identified in the first half of the paper. For instance, many authors concentrate on unproductive iteration combined with either process architecture related factors (*e.g.*, Browning and Eppinger, 2002) or management polices and project environment related issues (*e.g.*, Joglekar and Ford, 2005). The total impact of such strategies and policies on the wider product development system, which is influenced by the complex network of causes and effects indicated in Figure 8.1, has not been so widely studied. Thus, although existing approaches provide solutions which appear optimal from the perspectives they consider, they may not be robust under a broader consideration of potential influencing factors and impacts. For decision-makers it is essential to explore and understand both positive and negative effects of their actions as well as the total sensitivity of the process to the magnitude of those measures. We thus conclude that there is a need for a more integrated, holistic analytic approach to study iteration causes and effects. Such an approach should include both the network of causes and effects of iteration and the process architecture required to analyse the impact of iteration.

8.5 An Integrated Analysis Approach

The qualitative analysis and explanatory framework outlined above was used to develop an analytic framework which can be used to explore influencing factors of iteration, their interactions, the impacts of iteration on process behaviour, as well as possible causal loops and feedbacks among these entities. In overview, different process simulation frameworks can be combined to allow examination of the design process from different angles and abstraction levels, thus synthesising insights into different issues. This is based on the observation that detailed task-network process modelling frameworks - such as Design Structure Matrix (Browning and Eppinger, 2002) and Applied Signposting Model (ASM) (Wynn *et al.*, 2006) - capture process architecture related issues impacting on iteration and its management, while a high-level modelling framework - such as System Dynamics (SD) (Ford and Sterman, 1998) - is better-suited to explore the feedback structure of causes and effects and its impact on process behaviour.

Our integrated analysis framework uses a task network simulation model, such as an ASM model, as the starting point (see Figure 8.2). This model, which captures the architecture of a specific engineering design process, is then transformed automatically into the basic structure of an SD model. The process architecture is the core of the transformation since, as illustrated in the literature discussed in Section 8.4, process architecture is the main element which determines the impact of iteration on the performance of a specific process.

In an SD model, a PD process can be represented by work packages which flow through a standard 'rework cycle' (Ford and Sterman, 1998). The standard rework cycle model, depicted in Figure 8.2 operates as follows. At the beginning, all work packages are in the stock "Original Work to Do". They are processed at a given rate through "Work in Progress" until they reach "Work Done", where all work packages need to arrive to complete the process. However, some work packages will contain errors and need to be reworked once these errors are discovered. In this case, these work packages have to go from "Work in Progress" to "Undiscovered Rework" and "Rework to Do" before they can proceed to "Work Done". This rework cycle is further detailed by various feedback loops which govern the rates at which activities flow between the stocks described above. For instance, an increase in workload after rework has been discovered can have a negative impact on the productivity of work force due to limited resource, implying an increase in fatigue. This example captures one way in which resistance may evolve as management policies are initiated to influence the project.

In overview, our hybrid approach proceeds as follows. A process architecture is modelled using the ASM framework as described in Wynn *et al.* (2006), capturing individual tasks and their specific interdependencies. The ASM model is then simulated to determine a task schedule which could be visualised as a Gantt chart. Based on this task schedule, it is possible to calculate the percentage contribution of each successive time unit to the overall project scope, in terms of number of work packages completed. This information is fed into the construction of an SD rework cycle model as a process concurrency relationship, which limits the maximum number of work packages to be processed at each point in time. This basic structure of the rework cycle is then complemented with SD typical influencing variables and feedback structures developed from the network of iteration causes and effects summarised in Figure 8.1. Together with the process architecture, as summarised by the percentage contribution at each time unit, this causal network exerts strong influences on process behaviour. Modifying these factors and simulating the process can be used to develop a better understanding of how and to what extent each factor contributes to process performance.

Figures 8.2 and 8.3 illustrate our approach. The conceptual design process of a UK aerospace company was modelled through a series of interviews using a task-network approach (see Wynn *et al.*, 2006), then transformed into an SD model via the procedure outlined above (see Le *et al.*, 2010). Simulation of the SD model allowed the impacts of modifications to a process architecture-related factor which influences iteration (degree of activity overlapping) and a management policy-related factor (degree of schedule pressure) upon total process duration to be explored simultaneously. The results, shown in Figure 8.3, identify the non-linear effects of modifying these two factors in combination. This is intended as an illustrative example only; the strength of our approach is that it allows exploration of the effects of manipulating any combination of the levers from different process abstraction levels that affect iterative behaviour.

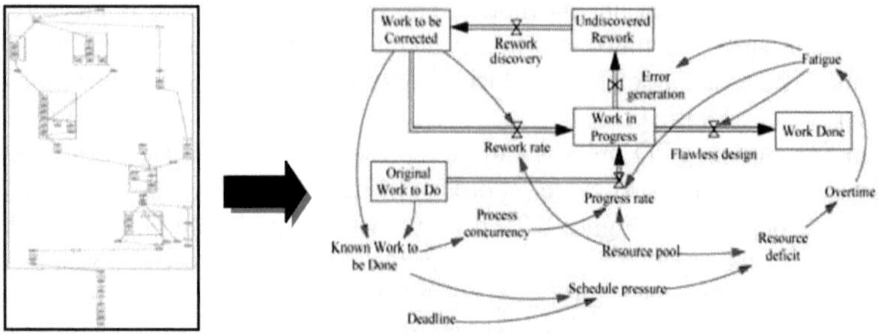

Figure 8.2. ASM basis model (*left*) and a simplified SD model showing a single rework cycle together with a subset of iteration causes and effects (*right*)

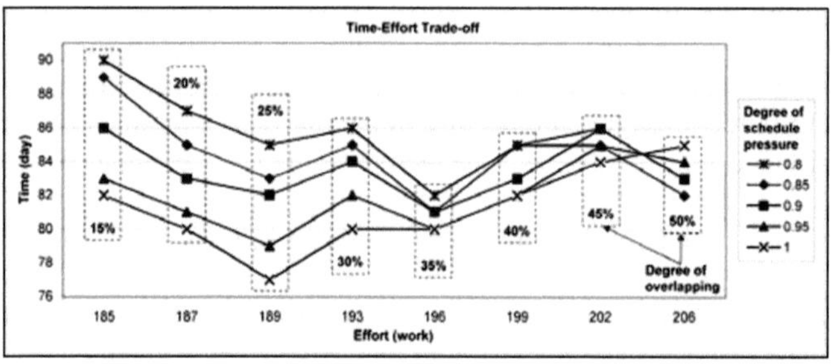

Figure 8.3. Application with varying degrees of process overlapping and schedule pressure

In summary, this hybrid simulation approach extends existing literature by allowing analysis of multiple factors which influence process performance from the two different abstraction levels of iteration causes and process architecture. Knowledge about the interactions between influencing factors is integrated. This is combined with a representation of the architecture of a specific process, which is required to quantify the ultimate impact of these influences in a particular case. The simulation approach provides a virtual environment offering a way to evaluate the impact of changes to process architecture and factors influencing iteration, considering that these changes can occur simultaneously and can accumulate in non-linear ways to give possibly unexpected outcomes.

8.6 Conclusions

Iteration is recognised to be unavoidable in the design process and many authors agree that it should be incorporated in some way when planning and managing projects in order to minimise surprises and reduce distortions of the project schedules. However, efficient planning and management of iteration under conflicting goals and constraints remains challenging. This paper has highlighted that many causes of iteration exist and that iteration may influence process behaviour directly and indirectly as well as positively and negatively. Some of the iteration causes were found to be partially under process management influence and control. This offers the opportunity to explore ways of optimising iteration management to reduce development time and cost. A review of analytic approaches which have been proposed to help manage iteration found that most consider only a relatively small number of iteration perspectives and influencing factors which were identified. Based on these findings, we developed an integrated analytic approach to study iteration causes and effects in combination. The approach builds on existing modelling approaches by facilitating a more holistic exploration of opportunities to reduce the negative impacts and enhance the positive effects of iteration. With further development, this approach could help to resolve some of the shortcomings of existing methods to help understand, and hence manage the complexity of iteration and its effects as encountered in practice.

8.7 References

Adams RS, Turns J, Atman CJ (2003) Educating effective engineering designers: The role of reflective practice. Elsevier Design Studies 24(3): 275-294

Adler PS, Mandelbaum A, Nguyen V, Schwerer E (1995) From project to process management - an empirically-based framework for analyzing product development time. Management Science 41(3): 458-484

Ahmadi R, Wang RH (1999) Managing development risk in product design processes. Operations Research 47(2): 235-246

Browning TR (1998) Modeling and analyzing cost, schedule, and performance in complex system product development. PhD Thesis, Massachusetts Institute of Technology, Boston, MA, US

Browning TR, Eppinger SD (2002) Modeling impacts of process architecture on cost and schedule risk in product development. IEEE Transactions on Engineering Management 49(4): 428-442

Cho SH, Eppinger SD (2005) A simulation-based process model for managing complex design projects. IEEE Transactions on Engineering Management, 52(3): 316-328

Clarkson PJ, Simons C, Eckert C (2000) Change propagation in the design of complex products. Design for Excellence: Engineering Design Conference 2000, John Wiley & Sons, Chichester, UK, pp 563-570

Cooper RG, Kleinschmidt EJ (1996) Winning business in product development: Critical success factors. Research Technology Management 39(4): 18-29

Costa R (2004) Productive iteration in student engineering design projects. MSc thesis, Montana State University, Bozeman, MT, US

Costa R, Sobek DK (2003) Iteration in engineering design: Inherent and unavoidable or product of choices made? In: Proceedings of the ASME International Design Engineering Technical Conferences (DETC'03), Chicago, IL, US

Denker S, Steward DV, Browning TR (2001) Planning concurrency and managing iteration in projects. Project Management Journal 32(3): 31-38

Eisenhardt KM, Tabrizi BN (1995) Accelerating adaptive processes - product innovation in the global computer industry. Administrative Science Quarterly 40(1): 84-110

Ford DN, Sterman JD (1998) Dynamic modeling of product development processes. System Dynamics Review 14(1): 31-68

Ford DN, Sterman JD (2003) The liar's club: Concealing rework in concurrent development. Concurrent Engineering-Research and Applications 11(3): 211-219

Fricke E, Schultz AP (2000) Coping with changes: Causes, findings, and strategies. Systems Engineering 8(4): 342-359

French M (1998) Conceptual design for engineers (3rd edn.). Springer-Verlag, London, UK

Gries M (2004) Methods for evaluating and covering the design space during early design development. Integration - the VLSI Journal 38(2): 131-183

Ha AY, Porteus EL (1995) Optimal timing of reviews in concurrent design for manufacturability. Management Science 41(9): 1431-1447

Hoedemaker GM, Blackburn JD, Van Wassenhove LN (1999) Limits to concurrency. Decision Sciences 30(1): 1-18

Joglekar NR, Yassine AA, Eppinger SD, Whitney DE (2001) Performance of coupled product development activities with a deadline. Management Science 47(12): 1605-1620

Joglekar NR, Ford DN (2005) Product development resources allocation with foresight. European Journal of Operational Research 160(1): 72-87

Kim YS, Cochran DS (2000) Reviewing TRIZ from the perspective of axiomatic design. Journal of Engineering Design 11(1): 79-94

Krishnan V, Eppinger SD, Whitney DE (1997) A model-based framework to overlap product development activities. Management Science 43(4): 437-451

Le HN, Wynn DC, Clarkson PJ (2010) Re-designing PD process architecture by transforming task network models into system dynamics models. In: Proceedings of the 11th International Design Conference (DESIGN 2010), Dubrovnik, Croatia

Loch CH, Terwiesch C (1998) Communication and uncertainty in concurrent engineering. Management Science 44(8): 1032-1048

Loch CH, Terwiesch C, Thomke S (2001) Parallel and sequential testing of design alternatives. Management Science 47(5): 663-678

Lyneis JM, Ford DN (2007) System dynamics applied to project management: A survey, assessment, and directions for future research. System Dynamics Review 23(2-3): 157-189

Mitchell VL, Nault BR (2007) Cooperative planning, uncertainty, and managerial control in concurrent design. Management Science 53(3): 375-389

Osborne SM (1993) Product development cycle time characterization through modeling of process iteration, MSc Thesis, Massachusetts Institute of Technology, Boston, MA, US

Pahl G, Beitz W (1996) Engineering Design - A Systematic Approach (2nd edn.). Springer, London, UK

Powell ALK, Mander C, Brown DS (1999) Strategies for lifecycle concurrency and iteration - a system dynamics approach. Journal of Systems and Software 46(2/3): 151

Pugh S (1996) Creatiing innovative products using total design. Addison-Wesley, Reading, UK

Roemer TA, Ahmadi R, Wang RH (2000) Time-cost trade-offs in overlapped product development. Operations Research 48(6): 858-865

Smith RP, Eppinger SD (1997) Identifying controlling features of engineering design iteration. Management Science 43(3): 276-293

Smith RP, Morrow JA (1999) Product development process modeling. Design Studies 20(3): 237-261

Sobek DK, Ward AC, Liker JK (1999) Toyota's principles of set-based concurrent engineering. Sloan Management Review 40(2): 67-83

Terwiesch C, Loch CH, De Meyer A (2002) Exchanging preliminary information in concurrent engineering: Alternative coordination strategies. Organization Science 13(4): 402-419

Wynn DC, Eckert CM, Clarkson PJ (2007) Modelling Iteration in Engineering Design. In: Proceedings of the 16th International Conference on Engineering Design (ICED'07), Paris, France

Wynn DC, Eckert CM, Clarkson PJ (2006) Applied signposting: A modeling framework to support design process improvement. In: Proceedings of the ASME International Design Engineering Technical Conferences (IDETC/CIE2006), Philadelphia, PA, US

Part III

Managing Product and Process Information

Chapter 9

An Operational Perspective for Capturing the Engineering Design Process

S. Gonnet, M.L. Roldán, H. Leone and G. Henning

9.1 Introduction

Most engineering design processes comprise knowledge-intensive and collaborative tasks. Design activities may involve many participants from various disciplines and require a team of designers and engineers with different skills to work together (Zha and Du, 2006; Wang *et al.*, 2009). Thus, teams of human experts, in conjunction with computer-aided tools, are the ones that by, interacting cooperatively, sharing resources of various types and intermediate design products (*i.e.* models that are generated, detailed specifications of resulting artefacts, drawings, sketches, *etc.*), solve intricate problems. Accordingly, several proposals have arisen to tackle design-related issues. Most of them try to address the multiple design data representation problem that has appeared due to the various design tools used by designers during a design process (ISO 10303-1, 1994; Marquardt and Nagl, 2004; Sudarsan *et al.*, 2005; Zha and Du, 2006). Furthermore, engineering data management systems (EDM) and product data management systems (PDM) provide assistance in managing products of a development process (*e.g.* models, diagrams, documentation files, *etc.*) along their life cycle (Chen *et al.*, 2008). As Westfechtel (1999) has pointed out, EDM and PDM systems focus on the products of development processes, neglecting the representation of the activities that have generated them. In fact, no effective and practical system exists for capturing, storing, and retrieving design knowledge and experience in collaborative product design activities (Zha and Du, 2006; Chen *et al.*, 2008). To overcome the above drawbacks suffered by current design support systems, this contribution proposes an operational-oriented approach to representing and capturing the engineering design process (DP). Thus, the design decisions are represented as design operations applied to the several versions of the products generated during the DP. Therefore, this proposal constitutes a means for (i) documenting the engineering DP, by capturing each executed operation while the design is carried out, and also (ii) maintaining the design history. Due to the increasing complexity and to the dynamic changes occurring in the engineering design domain, it is not possible to establish an "a priori" information model covering all the information items, work processes, and their relations that may be applicable in different projects and situations (Bayer and Marquardt, 2004). In fact, rather flexible and extensible frameworks for modeling design activities are needed (Nagl *et al.*, 2003). Therefore, the proposed model was designed in view of the requirement of supporting various design domains. It can be adapted to face a new engineering design problem by taking into account the particular concepts of the problem's domain and the possible operations that can be applied over the instances of those concepts. However, the design of chemical plants has been chosen to exemplify the proposed ideas.

The paper is organised as follows. In Section 9.2, the proposed model is described. It enables the capture of (i) activities, operations and actors that have generated each design product, (ii) the imposed requirements, as well as (iii) the rationale behind each adopted decision. Furthermore, it also offers an explicit mechanism to manage the different model versions that have participated during the DP. Afterwards, a prototype that was developed to validate our approach, named *TracED*, is presented in Section 9.3. This section describes how a particular engineering design domain can be defined in this environment, and then, by using it, how the several products of a DP can be captured and traced.

9.2 Representing the Engineering Design Process

Unfortunately, once a design project is finished, those things that remain are mainly design products (*e.g.* generated models, detailed specifications of resulting artefacts, drawing, sketches, *etc.*). However, there is no explicit representation of how they were obtained. More specifically, there is no trace of which activities originated a given product, which requirements were imposed, which actors performed a given activity, and what the underlying rationale behind a decision-making activity was. The class diagram presented in Figure 9.1 introduces the main concepts used by this approach to model the engineering DP. The proposed model considers a DP as a sequence of activities that operate on the products of the DP, called *design objects* (Figure 9.1). *Activities* are the tasks that are carried out during design processes. In the chemical engineering (ChE) domain, when tackling the design of a separation system, typical activities are: proposing a given separation structure, analysing whether such structure satisfies the imposed separation targets, evaluating its economic potential, deciding on alternative separation schemes, *etc.* They may be described at various abstraction levels; thus, an *activity* may be decomposed into a set of *sub-activities* (Figure 9.1). The realisation of one activity is guided by one or more *requirements* (Figure 9.1). Therefore, the engineering DP is interpreted as a series of activities guided by requirements that specify the functional and non-functional characteristics that a product must satisfy. *Activities* are performed by *actors*, and an *actor* may be either an *individual* (a human being or a computational program) or a *team* (Figure 9.1). Teams allow representing compound *skills* that are needed for performing activities. Each *basic activity* performed by an *actor* during a DP is represented through the execution of a sequence of *operations*, which transforms the *design objects* (Figure 9.1). Typical design objects are models of the *artefact* being designed, specifications to be met (stream purity specs, products' throughput, *etc.*), variable values (*e.g.* reflux ratios, number of stages of a separation unit, operating temperatures and pressures, etc). Naturally, these objects evolve as the design process takes place, giving rise to several versions.

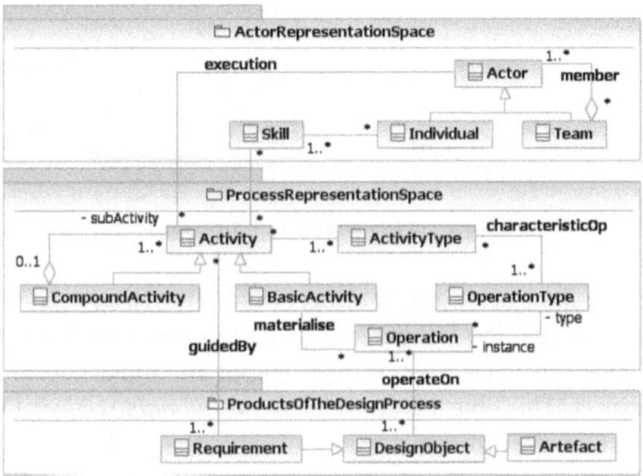

Figure 9.1. Engineering design process model

Versions may be considered as snapshots taken to *design objects* at a given time point, and a set of versions comprise a *model version*. A model version describes the state of the DP at a given time, including the *artefact* being designed. For example, Figure 9.2 partially shows two model versions, m_k and m_q. Both are the result of a fragment of the design of a chemical plant that must produce 40000 tons of polyamide-6 per year (Nagl *et al.*, 2003). Model version m_k represents the state of the engineering DP after the literature study has been carried out and the decision between batch and continuous operation has been made. A block flow diagram representing the continuous polymerisation of polyamide-6 is shown in the partial view of m_q. In this scheme, each *model version* is generated by applying a *basic activity* on a *predecessor model version*. The *basic activity* is *materialised* by a sequence of *operations* (ϕ on Figure 9.2) that may include the elimination, creation, and modification of versions that constitute the predecessor *model version*. Therefore, all model versions may have zero or more successors and must have only one predecessor model version (except the initial one). Consequently, the representation scheme of versions is a tree structure, where each model version is a node and the root is the initial model version. Figure 9.2 illustrates the evolution from model version m_k to m_q. This last one is obtained from the model version m_k by performing the flowsheet synthesis basic activity that has revealed that there are three main parts of the process to be designed: reaction, separation and extrusion. This activity is materialised by deleting *Polyamide-6Section* and adding the following sections: *ReactorSection*, *SeparationSection*, and *ExtrusionSection*, with their ports and flows (Fig. 2).

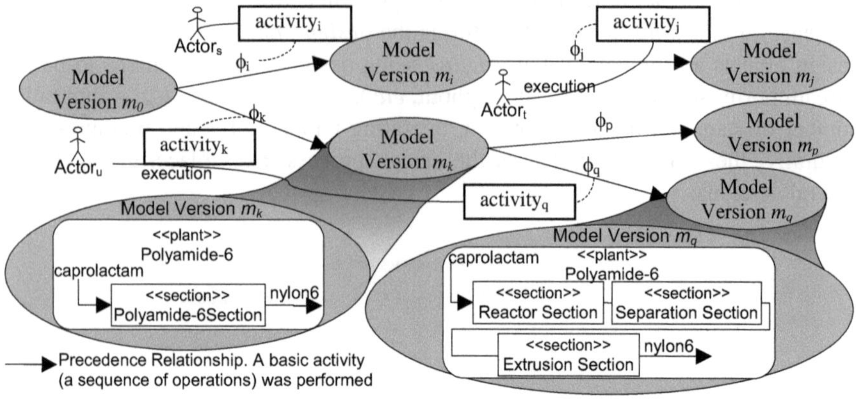

Figure 9.2. Tree-structure representation of an engineering design process

In order to maintain these versions, a *design object* is represented at two levels, the *repository* and the *versions' level* (Figure 9.3). The repository level keeps a unique entity for each design object that has been created and/or modified due to model evolution during a design project. This object is considered a *versionable object.* On the other hand, the *versions' level* keeps the different versions of each design object, which are referred to as *object versions*. The relationship between a versionable object and one of its object versions is represented by the *version* relationship (Figure 9.3). Therefore, a design object keeps a unique instance in the *repository* and all the versions it assumes belong to the *versions' level.* The class diagram illustrated in Figure 9.3 represents the main concepts of the version administration model. At a given stage during the execution of a design project, the states assumed by the set of relevant design objects comprise a *model version*, which supplies a snapshot description of the state of the engineering DP at that point. Each transformation operation that is applied to a model version incorporates the necessary information to trace the model evolution. This information is represented by the *operation* concept that relates the object versions to which the operation is applied (*argument*) and the ones arising as the result of its execution (*result*, Figure 9.3).

When capturing the versions generated during a DP, a design support computational tool has to be generic and flexible to allow the definition of *design object types* that are specific to the engineering domain being tackled. With this purpose, the *Domain* package, shown in Figure 9.3, enables the definition of the needed concepts. This package allows the specification of a particular domain, where the structure of the design objects is detailed. In fact, design objects are the building blocks employed to represent design artefacts, requirements, and other design products, such as the arguments of the adopted decisions. For each design object type, whose history has to be maintained, an instance of *DesignObjectType* is generated and its versionable properties are specified by a set of instances of the *Property* class. Furthermore, the possible relationships among these concepts will be instances of the *DomainRelationship* class of the *Domain* package. In Figure 9.4, two partial views of a domain model are illustrated.

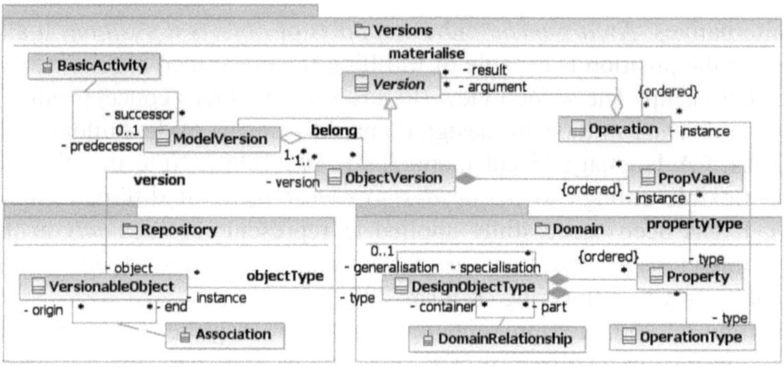

Figure 9.3. Version model that captures and traces the engineering design processes

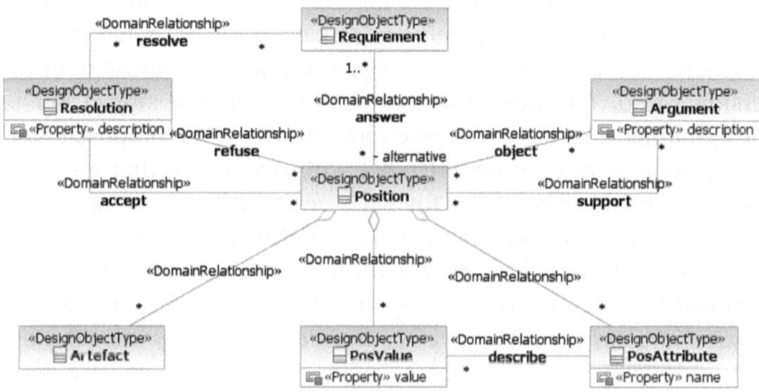

(a)

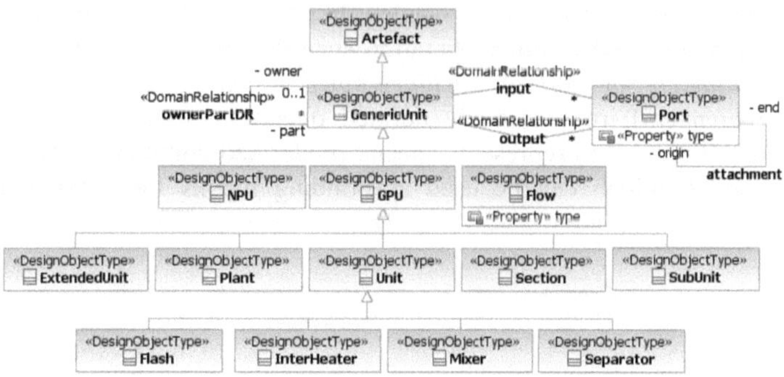

(b)

Figure 9.4. A domain model capable of representing: (a) IBIS concepts; (b) artefact related concepts in the ChE domain by means of the MODEL.LA language

The first view (Figure 9.4a) includes the relevant concepts for representing the rationale associated with the execution of a given design activity. In this case, the IBIS related (Kunz and Rittel, 1970) concepts have been chosen. Therefore, the different *alternative* products that arise in the DP are represented by the *position* concept. A *position* is qualified by one or more *arguments* and addresses at least one *requirement*. Attributes (*PosAttribute*) and values (*PosValue*) characterise the design alternatives. An *argument* either *supports* or *objects* a *position*. It allows to test whether the position is capable of fulfilling the prescribed requirements by the answer relationship. The second view (Figure 9.4b) includes concepts employed to represent an *artefact* in the ChE design domain by means of the building blocks of the MODEL.LA language (Stephanopoulos *et al.*, 1990). Note that the proposed model is flexible enough to define *design object types* of different engineering domains. It has been successfully applied to represent *requirements* in the ChE domain (Gonnet *et al.*, 2007) and *artefacts*, *requirements*, and *design rationale* in the area of software engineering (Roldán, 2009).

The proposed approach provides mechanisms to capture the several products generated during an engineering DP, as well as the elements for specifying, executing, and capturing the operations that generated them. Gonnet *et al.* (2007) have proposed the primitive operations *add*, *delete*, and *modify* to represent the transformation of model versions. By using the *add* operation, an object version that did not exist in a previous model version can be incorporated into a successor one. Conversely, *delete* eliminates an object version that existed in a previous model version. Also, the *modify* operation creates a new object version of an existing design object. However, to give more power to the proposed model, these primitive operations can now be extended with operations that are suitable to a specific design domain. Therefore, for each *design object type* a set of possible *operations* (*OperationType* in Figure 9.3) is defined. The model introduced in Figure 9.5 enables the specification of *operation types* and their implementation in a computational tool. An *operation* (Figure 9.5) is defined as a macro command (*MacroCommand* class), a subclass of *command* that simply executes a sequence of commands. In consequence, when an operation is specified, it is necessary to define both the *arguments* and the *body* of the *operation*. The *body* of a *MacroCommand* is comprised by some already defined *commands* that are available for being used in other operation specifications. They can be *primitive* (such as *add*, *delete*, or *modify*), *auxiliary function* commands (like *iteration* and *variable assignment*), or previously defined operations. The *AuxiliaryFunction* abstract class (Figure 9.5) represents special predefined commands offered by the operation model, which can be used as building blocks for defining complex operations.

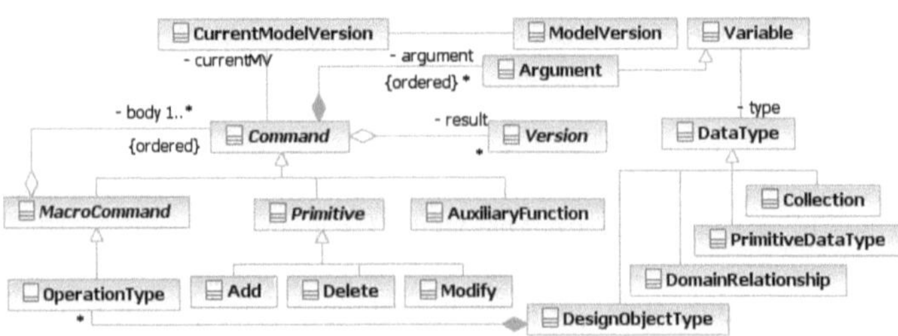

Figure 9.5. Operations package

As shown in Figure 9.5, every command has one or more *Arguments*. An *Argument* is considered as a kind of *variable*. A *variable* can also be declared and used in the body of an operation and has a given type. The abstract class *DataType* generalises the types described in the model; so, *PrimitiveDataType* and *Collection* are subclasses of *DataType*. *PrimitiveDataType* includes the integer, float, string and boolean types, and *Collection* describes an ordered list of elements of a particular type. Furthermore, *DesignObjectType* is included as another data type. For example, Figure 9.6 presents the functional specifications of some basic design operations in the ChE design domain. It is important to note that each functional specification of an operation proposed in this section implies the instantiation of the operation model (Figure 9.5).

```
addGPU(o:GPU, gpuname:String,
    l_InputPorts:Collection[String],
    l_OutputPorts:Collection[String],
    l_props:Collection[PrimitiveDataType])
    gpu:= add(gpuname, GPU, l_props)
    for each p in l_InputPorts
        addInputPort(gpu, p, [])
    end for
    for each p in l_OutputPorts
        addOutputPort(gpu, p, [])
    end for
    addRelationship(o, gpu, ownerPartDR)
    end

addFlash(o:GPU, fname:String,
    l_InputPorts:Collection[String],
    l_OutputPorts:Collection[String],
    l_props:Collection[PrimitiveDataType])
    f:= add(fname, Flash, l_props)
    for each p in l_InputPorts
        addInputPort(f, p, [])
    end for
    for each p in l_OutputPorts
        addOutputPort(f, p, [])
    end for
    addRelationship(o, f, ownerPartDR)
end

addFlow(o:GPU, fname:String, l_InputPorts:Collection[String], l_OutputPorts:Collection[String],
    l_AtInputPorts:Collection[String], l_AtOutputPorts:Collection[String], l_props:Collection[PrimitiveDataType])
    f:= add(fname, Flow, l_props)
    i:= 0
    for each p in l_InputPorts
        addInputPort(f, p, [])
        addRelationship(l_AtInputPorts(i), p, attachment)
    end for
    for each p in l_OutputPorts
        addOutputPort(f, p, [])
        addRelationship(l_AtOutputPorts(i), p, attachment)
    end for
    addRelationship(o, f, ownerPartDR)
end

addInputPort(g:GPU, pname: String,
    l_props:Collection[PrimitiveDataType])
    p:= add(pname, Port, l_props)
    addRelationship(g, p, input)
end

addOutputPort(g:GPU, pname: String,
    l_props:Collection[PrimitiveDataType])
    p:= add(pname, Port, l_props)
    addRelationship(g, p, output)
end

deleteFlash(f: Flash)
    l_Ports = get(Port, f)
    for each p in l_Ports
        deletePort(p)
    end for
    delete(c)
end

deletePort(p: Port)
    delete(p)
end
```

Figure 9.6. Functional specification of certain design operations in the ChE domain

Any design operation is defined in terms of primitive operations, like *add(name, DesignObjectType, listOfPropertyValues)*, and non-primitive ones, such as *deletePort(p: Port)*. For example, the *addFlash(o:GPU, fname:String,* $l_{InputPorts}$*:Collection[String],*$l_{OutputPorts}$*:Collection[String],*l_{props}*:Collection[Primitive DataType])* operation allows adding a *flash unit* to a generic processing unit (*GPU*) *o*. As it can be seen in Figure 9.6, this operation is carried out by a series of simpler

operations. First, a version of a *flash* unit *f* is added ($f := add(fname, Flash, l_{props})$), where the last argument is an ordered list of property values, such as temperature, pressure, *etc.* After that, the sets of input and output ports (detailed by the $l_{InputPorts}$ and $l_{OutputPorts}$ list arguments) are inserted. It must be remarked that the addition of a new element into a model version means the instantiation of a certain *design object type*. It also implies the creation of a *versionable object* at the repository level and the generation of an *object version* at the version level. Some of these design objects can also have *associations* at the repository level. Thus, in the functional specification of operations, the pseudo-operation *addRelationship* is used, which adds an association between two versionable objects into the repository.
Functional specifications give an outline of how the operations might be configured using a computational tool. They are useful since they provide the syntax on which a computational tool that realises the proposed model is to be based. Section 9.3 introduces an example of how the *TracED* prototype, which implements the proposed model, allows the definition of operations in the ChE domain.

Just as domain-specific basic operations can be specified, it is also possible to define rationale-related operations, such as *addAlternative* or *addSupportingArgument* (Figure 9.7), or more abstract design operations that refine a design element or apply a design decision, in the same way a certain design pattern is applied in software architecture design. Figure 9.8 includes a refinement operation, named *refineGPU.* This operation refines a generic processing unit into several generic processing units with their ports and flows. This operation was applied in the case illustrated in Figure 9.2. Thus, the design object *Polyamide-6Section* was refined into *ReactorSection*, *SeparationSection*, and *ExtrusionSection*, the flows among these new sections were added, and the flows associated with the ports of the *Polyamide6-Section* (named *caprolactam* and *nylon6*) were also attached to the new sections.

```
addAlternative(r: Requirement, pName: String,
    lprops:Collection[PrimitiveDataType])
    p := add(pName, Position, lprops)
    addRelationship(r, p, answer)
end

addSupportingArgument(argName: String,
    p:Position, lprops:Collection[PrimitiveDataType])
    a := add(argName, Argument, lprops)
    addRelationship(a, p, support)
end
```

Figure 9.7. Functional specification of design rational-related operations

9.3 TracED

TracED is a prototype devised for validating the proposed model, which was conceived for capturing and tracing engineering designs. It has been developed using the Java language, and the MySQL database. The Hibernate framework has been employed in order to achieve object persistency on the relational database. The major components of *TracED* are the *Domain Editor* and *Versions Manager*. By making use of the *Domain Editor*, a domain expert can specify one or more engineering design domains. Thus, the *Domain Editor* allows the definition of *design object types* along with their suitable *operations*. Then, by working with the *Versions Manager*, designers can choose an existent domain and use the design object types and operations already defined in it to carry on design projects.

```
refineGPU(g: GPU, l_gpus:Collection[String], l_lPorts:Collection[Collection[String]], l_Flows:Collection[String],
    l_InputFlowPorts:Collection[Collection[String]], l_OutputFlowPorts:Collection[Collection[String]],
    l_lprops:Collection[Collection[PrimitiveDataType]])
    i := 0
    for each gpuname in l_gpus
        agpu := addGPU(get(ownerPartDR, g), gpuname, l_lPorts(i), l_lprops(i))
        i++
    end for
    i := 0
    for each nflow in l_Flows
        l_inputp := select(get(Port, null)) // to select ports to attach the added flow
        l_outputp := select(get(Port, null)) // to select ports to attach the added flow
        addFlow(get(ownerPartDR, g), nflow, l_InputFlowPorts(i), l_OutputFlowPorts(i), l_inputp, l_outputp, [])
        i++
    end for
    lp := get(Port, g)   //to get the ports of the refined GPU
    for each p in lp
        r := get(Port, p)  //r is the port attached with p
        np := select(get(Port, null))          // select is a interactive function,
                                               // asks the designer the port to be attached to r
        addAssociation(np, r, attachment)
    end for
    deleteGPU(g)
end
```

Figure 9.8. Functional specification of a particular refine design operation

The *Domain Editor* implements the domain (Figure 9.3) and operation (Figure 9.5) packages. Therefore, the *design object types* which were presented in Figure 9.4 can be defined. A partial view of the MODEL.LA concepts specified in *TracED* is shown in Figure 9.9. Furthermore, the *Domain Editor* provides features for specifying operations; thus, Figure 9.9 shows the definition of the *addFlash* operation, which follows the functional specification that was presented in Figure 9.6.

The *Versions Manager* enables the execution of a design project. When a new design project is created, an existing *design domain* has to be selected. Therefore, the evolution of a project is based on the execution of domain-specific *operations* and the instantiation of the *design object types* pertaining to the selected *design domain*. Additionally, *TracED* includes in its *Versions Manager* features which allow keeping information about: (i) predecessor and successor model versions (if any exists) of each model version; (ii) history links which save traces of the applied operation sequences (basic activities), which originated new model versions; (iii) references to the set of object versions that arose as a result of each operation execution. Figure 9.10 illustrates the *TracED* implementation of the model versions that have already been presented in Figure 9.2. The figure exemplifies the execution of *refineGPU* operation, which refines the *Polyamide-6Section* object version.

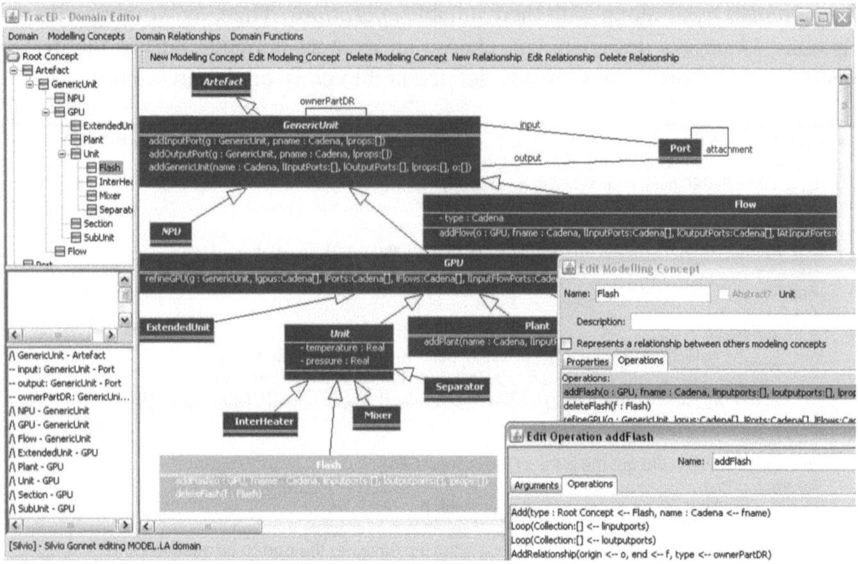

Figure 9.9. Partial view of elements of the ChE design domain using the MODEL.LA language

In almost all engineering domains there is a real need for recalling how an engineering design project has evolved. As it was previously pointed out, an instance of the *Operation* class is created for each executed operation. Thus, it is possible to reconstruct the history of a given model version by beginning the trace from the initial one. Figure 9.10 shows an example of how *TracED* depicts history-related information. It presents the history window, which informs all the operations that have been applied to evolve from the initial model version to the selected one (the figure only shows *ModelVersion5* and *ModelVersion6*). In this window, it is possible to see detailed data about each applied operation. For instance, it presents information about the time point at which the operation was applied, who the actor involved was, and the names assigned to the new object versions (*successor object versions*). In this case, the history window shows that a *refineGPU* operation was executed at the top of *ModelVersion6*.

The history window is a tool for presenting the records of a DP. On top of it, the model allows generating several queries to exploit the information already caught. By having captured a DP, the model can answer questions like the following: (i) by means of which design operations a particular model version was obtained?; (ii) which are the alternative successor model versions of a certain model version?; (iii) how did a design object change along the DP?; (iv) who were the actors involved in the generation of an object version?; (v) which were the products generated due to the execution of a given operation in a basic activity?

It is important to note that *TracED* was developed with the aim of just showing that it is possible to apply the proposed model for capturing engineering design processes, but it is not intended to replace traditional design environments. On the contrary, *TracED* should be integrated with existing tools. In this way, *TracED* would perform the capture of all applied operations, by working in a background mode, without being noticed by the designers.

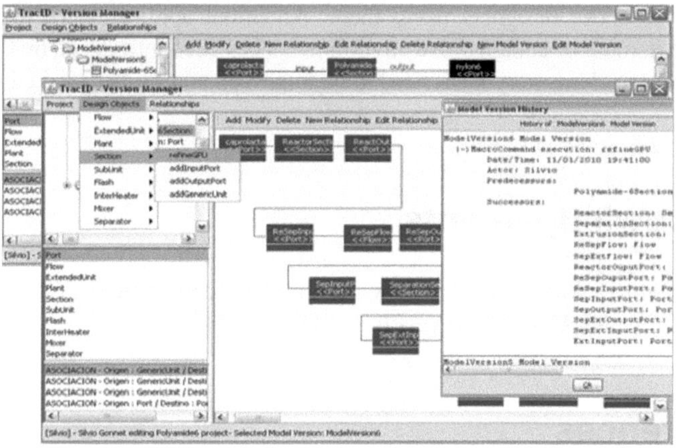

Figure 9.10. Version Manager. Partial view of *ModelVersion6* after *refineGPU* execution

9.4 Conclusions

An operational-oriented approach to capturing and tracing engineering design processes has been presented. It naturally integrates the representation of the products of the DP with the activities that have generated them, capturing also the participating actors and their rationale along the DP. Due to these capabilities, the history of a design project can be naturally captured and stored in an efficient way, allowing a latter retrieval of knowledge and experience.

The underlying ideas have been illustrated within the ChE domain by means of *TracED*, a research prototype which was devised for validating the proposed framework. The approach provides mechanisms for capturing the several products generated during a DP, as well as the elements for specifying, executing, and capturing the operations that have originated them. The proposal is general and can be applied to various engineering domains. In the software engineering arena, Roldán (2009) has illustrated how the proposed model supports the specification of design operations that apply software design patterns. Such an application refines a software component instance with a new set of components and connectors that are instantiated from a pre-existing design pattern. This process is not fully automatic and requires the designer participation to define how to delegate responsibilities and scenarios (requirements), as well as how to map connectors between external components and refined components.

In addition, this contribution has shown that the elements of a particular design domain, and the operations that are suitable for it, can be specifically defined. Based on this feature, a future extension of the presented approach will consist in developing tools for supporting conflict management processes and building features for taking care of conflict detection and resolution. As expected, these tools are not going to be off-line ex-post features of a design kit. Quite the opposite, they need to be available on-line, giving support to a collaborative design team. As well as with the other features of the TracED environment, the idea is to have all these tools working in a background mode behind the traditional computed-aided design support tools, without being noticed by those designers using them. An initial application of the present proposal in the domain of collaborative environments has been introduced by Gonnet *et al.* (2007), but these elementary ideas need to be improved and further worked on in the future.

9.5 Acknowledgements

The authors wish to acknowledge the financial support received from CONICET (PIP 2754), UNL, UTN and ANPCyT (PAE-PICT-2007-02315).

9.6 References

Bayer B, Marquardt W (2004) Towards integrated information models for data and documents. Computers and Chemical Engineering 28(8): 1249-1266

Chen YJ, Chen YM, Chu HC (2008) Enabling collaborative product design through distributed engineering knowledge management. Computers in Industry 59(4): 395-409

Gonnet S, Leone H, Henning G (2007) A model for capturing and representing the engineering design process. Expert Systems with Applications 33(4): 881-902

ISO 10303-1 (1994) Industrial automation systems and integration - product data representation and exchange - Part 1: Overview and fundamental principles. International Organization for Standardization (ISO), Geneva, Switzerland

Kunz W, Rittel HWJ (1970) Issues as elements of information systems. Working Paper 131, Institute of Urban and Regional Development, University of California, Berkeley, CA, US

Marquardt W, Nagl M (2004) Workflow and information centered support of design processes - the IMPROVE perspective. Computers and Chemical Engineering 29(1): 65-82

Nagl M, Westfechtel B, Schneider R (2003) Tool support for the management of design process in chemical engineering. Computers and Chemical Engineering 27: 175-197

Roldán ML (2009) Un modelo para la representación de conocimiento y razonamiento en el proceso de diseño basado en arquitecturas de software. PhD Thesis, Facultad Regional Santa Fe, Universidad Tecnológica Nacional, Santa Fe, Argentina

Stephanopoulos G, Henning G, Leone H (1990) MODEL.LA. A modeling language for process engineering: Part I. The formal framework. Computers and Chemical Engineering 14(8): 813-846

Sudarsan R, Fenves SJ, Sriram RD, Wang F (2005) A product information modelling framework for product lifecycle management. Computer-Aided Design 37: 1399-1411

Wang J, Tang M, Song L, Jiang S (2009) Design and implementation of an agent-based collaborative product design system. Computers in industry 60(7): 520-535

Westfechtel B (1999) Models and tools for managing development process. Springer-Verlag, Berlin, Germany

Zha X, Du H (2006) Knowledge-intensive collaborative design modeling and support. Part I: Review, distributed models and framework. Computers in Industry 57(1): 39-55

Chapter 10

Sparse Data Estimation in Complex Design Processes

K. Lari, O. Hisarciklilar, K. Grebici and V. Thomson

10.1 Introduction

Process monitoring is one of the major components in any process management system. A process monitoring system is an essential set of tools for aiding individual(s) responsible for the process to understand and analyse performance, potential problems, and help continuous improvement of processes. An integrated process monitoring system is a combined tool kit in which various elements of a process monitoring system automatically interact with each other within a common framework. This reduces the necessity for human interaction, while improving information flow among various components of such a system. There have been efforts to design process control and monitoring systems; however, no integrated system has yet been developed as a "generic intelligent system shell" (Karsai *et al.*, 1992).

Furthermore, a comprehensive study of the available methodologies and techniques revealed that sparse data estimation (SDE) is the key component of an Integrated Process Monitoring System (IPMS), which aims to provide a total end-to-end solution for process monitoring (Lari, 2004). SDE is especially crucial, yet difficult to do in the context of knowledge processes, such as design, which are characterised by a high level of uncertainty. The unknown or partially known (sparse) information often results in uncertain cost and duration. Techniques that allow the estimation of sparse data greatly improve management capabilities over design processes, especially by providing continuous estimation of data throughout the process and automating process monitoring. This can add significant advantages towards reducing cost, duration and risk.

Consequently, this paper presents a novel architecture for an Integrated Process Monitoring System along with a Sparse Data Estimation method. A series of algorithms, as well as several artificial intelligence (AI) techniques are incorporated into the IPMS architecture to address sparse data estimation for design processes.

The remainder of this paper consists of the following sections: Section 10.2 provides an overview of existing process modelling, analysis and estimation techniques based on mathematical as well as artificial intelligence techniques suitable for a process monitoring system. In Section 3, a novel architecture of an IPMS is introduced. Section 4 focuses on the SDE component of the IPMS. The application of the SDE algorithms on a design process example is presented in Section 5. Analysis and insights conclude this section. Finally, the conclusion and future suggested research areas are presented in Section 10.6.

10.2 Background Research

This section presents the methods used in a proposed monitoring system that could be applied to any knowledge intensive process (*e.g.*, design). This monitoring system includes process modelling, process analysis and data estimation.

10.2.1 Process Modelling Methods

SADT (System Analysis and Design Technique) (Ross and Schoman, 1977) and later on Integrated Definition methods (IDEF) have been used to represent information flows and knowledge processes. These modelling methodologies are of importance due to their capability to facilitate analysis. The IDEF3 methodology is of particular interest since it is used to model processes (Mayer *et al.*, 1992). Indeed, IDEF3 has the advantage of being able to provide the description of activities as well as timing and task sequencing. These features comply with the need to analyse and compare different process designs. SADT and IDEF allow the capture of the causality relations within processes.

Certain methods seek to simulate the flow of information throughout a process, and in particular, the flow during work iteration (Wynn, 2007). These latter methods allow the study of the iterative dynamics in processes.

10.2.2 Process Analysis Techniques

10.2.2.1. Critical Path Method (CPM)

The interest in CPM is the trade-off that is offered between the cost of a project and its overall completion time. CPM is an operational research algorithm for finding the longest path(s) through a set of activities (Hebert, 1975). CPM facilitates the analysis of processes or projects represented as network diagrams. The approach requires the consideration of 1) what activities are to be done, 2) the sequence in which they are performed, 3) the resources required, and 4) the time taken by each activity. A set of graphed vertices and connecting arcs represent the process. The vertices stand for events, and the arcs stand for the activities of a process. The critical path is the path from the start event to the end event which takes the longest time or where slack time is zero.

Other notable techniques for planning and controlling a process are PERT (Performance Evaluation and Review Technique) and GERT (Graphical Evaluation and Review Technique); they determine project or process completion in the shortest possible time (Malcolm *et al.*, 1959). The critical path includes a sequence of activities that cannot be delayed without jeopardising completion of the entire process. PERT can be used to estimate the probability of completing either a process or individual activities by any specified time. These techniques consider a project to be an acyclic network of events and activities; they also ignore the multi-alternative characteristics of processes. In this study, an extension of CPM is considered in conjunction with an IDEF3 process modelling technique.

10.2.2.2 Process Structure Analysis

Other interests regarding process analysis are considered with regard to the structural importance that tasks or process steps play within a complex process network. The factors of process complexity considered here are the presence of repeated sub-trees and repeated primary tasks, and the set of alternative processes (including excusive process routes (where tasks are related by an XOR type of link) or alternative non-exclusive alternatives (OR links)).

Triangularisation, fault tree, structural importance, minimal cut set, and minimal path set algorithms are appropriate techniques that were used by Larson and Kusiak (1996) to calculate the structural importance of the tasks using an IDEF3 process model. In fact, Larson and Kusiak considered the structure of a complex process as a fault-tree model where the parent activity consisted of the final task in a sequence of tasks. The success or the reliability of the parent task relied on the success or the reliability of its children. The higher the value of the measure of structural importance, the more important was the parent activity of the model from a structural point of view.

10.2.3 Mathematical and Statistical Methods for Data Estimation

10.2.3.1 Sparse Data

For the purpose of this research, the term *sparse data* is chosen to represent any situation, environment, circumstance, event, *etc.*, where the entire data collection is not known; in other words, a portion of the data is missing. For simplicity, the term *sparse data* refers to both data and information. Sparse data occurs in almost every data collection event. This is due to many factors such as, time, cost, complexity, data collection methods, circumstances, observers' sensitivity, hardware errors, human errors, organisation, knowledge, *etc.* In this paper, particular attention is drawn to the estimation of task duration and process event times.

There are different ways of assessing missing data. We focus on the *missing at random* case (Little and Rubin, 1987). The *missing at random* corresponds to the situation where the missing element can be assessed based on the observed data. For instance, people may answer readily certain questions about their lifestyle, but may be less inclined to give details about their income. Their income can be estimated, however, through other observed descriptors such as, their education level.

10.2.3.2 Missing Data Estimation Techniques

Many estimation methods of missing data can be considered as maximising the likelihood function under certain modelling assumptions. For instance, a general-purpose algorithm for maximum likelihood estimation in a variety of problems including missing data problems is proposed by Dempster *et al.* (1997).

Mathematical estimation techniques usually have a significant amount of calculation with a considerable level of complexity even for relatively simple sparse data problems. They also lack the desired flexibility required when facing real life situations, in which mathematical modelling of processes is very troublesome or conceivably impossible.

Knowledge-Based (KB) and Artificial Intelligence (AI) methods are the most cited and used methods for data estimation. In KB the information repository may contain different kinds of knowledge with different formats. A knowledge base can be called *rule-based* when the information in the KB is represented as a set of rules defined by experts usually in an if-then-else format. Rule-based systems are fairly simplistic, but provide the basis for so-called 'expert systems'.

Ordinary ruled-based KBs are only capable of interpreting information as delimited by the input condition(s) of their rule base, and cannot create new rules (Turksen and Zhao, 1993). Therefore, these systems usually fail abruptly when the environment changes and new conditions are not represented in the KB as experienced knowledge (Lenat and Guha, 1990). Rule-based KBs are also information intensive systems and require a considerable amount of data to build them. Fuzzy logic and fuzzy methods have been widely employed in KBs. The fuzziness can be applied to each piece of information, the inference logic, and interrelations among individual rules (Turksen and Tian, 1995).

This hybrid intelligent system including rule-based KB and fuzzy logic has been selected for the data estimation and the validation step of the Sparse Data Estimation (SDE) algorithm (Section 10.4.1).

10.3 Integrated Process Monitoring System

An architecture of an integrated process monitoring system (IPMS) is presented in this section. The objective of such an architecture is to address industry requirements in the area of design processes, which can provide a total end-to-end solution for process monitoring, *i.e.*, process identification, process modelling, process measurement, process comparison, process analysis, progress status, fault diagnosis, report generation, *etc.*

An IPMS compares a real world process model to a desired process model. 8 main components are proposed by Lari (2004) to provide a total end-to-end solution for process monitoring:

- **process modeller**, which facilitates the design and planning of processes.

- **metric analysis system (MAS)**, which determines the required data to be measured. The IPMS utilises several techniques, similar to CPM or PERT, to determine essential processes for the measurement of each set of metrics.
- **sparse data estimator (SDE)** which estimates and normalises data obtained from: 1) the real world, 2) the knowledge repository, and 3) simulation results.
- **real time process comparison engine (PCE)**, which compares the performance of a real world process to a model of the desired process. A comparison algorithm is proposed by Nayestani (2002). The comparison is performed on a peer-peer basis; therefore, detailed comparative results are available for the different levels of decision making.
- **fault diagnoser (FD)**, which determines possible area(s) of faults and their causes, and consequent effects; diagnoses faults and failures, or any deviation in the real world system from the normal operating environment is determined by simulation.
- **feedback generator (FBG),** which delivers reports on the performance of processes;
- **knowledge base (KB)**, which contains model objects, best practices, historical data, and algorithms and procedures for measuring the characteristics of processes; and
- **simulator**, which simulates real processes for process review as part of feedback.

It has been realised that sparse data estimation is the most critical yet least studied component of IPMS. Accordingly, the next sections describe a sparse data estimation model as a part of an IPMS.

10.4 Sparse Data Estimation

Although there is a significant amount of research and numerous methods to apply to incomplete data in order to estimate the missing parameters of the observed environment, these methods are efficient in the presence of a large amount of data. However, this is not the case when approaching knowledge-intensive processes. In fact, although there are many processes and a large number of process characteristics, there is usually a very limited amount of performance data available at any given time in order to apply traditional mathematical and statistical techniques.

As mentioned earlier, an IPMS compares two sets of information. The first set of information is gathered from the real world system. Real world data is made ready for comparison through a normalisation process. During normalisation, the missing pieces of information are estimated and the gaps are filled. Once the information reaches the comparison engine, it must be complete. To allow comparison, it is imperative that the data sets of both the ideal and the real world process models be identically populated.

The second set of data is derived from the ideal process model. An ideal process model is a model constructed from an abstraction of the real world process. The architecture of the ideal model is identical to the actual process. However, it differs from real life process models by the methods that determine the values for various attributes of its processes. In a real life process model the values are collected from actual events in a single process, whereas in an ideal process model the values are computed using simulation and other mathematical methods (Section 2.3.2) or are averages of values collected from actual processes over time.

The intention for sparse data estimation is to finish populating a real world process model with data; so, the real world process can be compared to an ideal model of the same process.

10.4.1 Key components for SDE

The key components for sparse data estimation are presented below. These components are used at different stages of a sparse data estimation procedure for information gathering, analysis and validation. In the next section, an algorithm for sparse data estimation is proposed and the roles of these components throughout the SDE process are discussed.

Process simulation

Process simulation is undertaken to estimate incomplete data collected from the real world. The simulator uses a validated model and all related information, *i.e.*, resource availability, resource constraints, *etc.* to produce simulation results using a desired model. To estimate incomplete data, for each estimation phase, numerous iterations are performed. The results of these operations are two sets of comparable model information, one set from the desired (ideal) model and another set from the real world system.

Process analysis

Process analysis includes methods for network junction simplification and application in IDEF3 methodology. The IDEF3 model is considered as a basis upon which the application of algorithms for task structural importance (including the path finder algorithm) is considered for a wide range of analyses such as critical path, reliability, fault tree, *etc.*

Knowledge base

Constraints related to a specific metric that exist in a constructed knowledge base are consulted in order to verify the validity of a new set of data. Each of the rules and facts related to the estimated data is verified within the knowledge base. These rules and facts can for instance result from the process structure, or process attributes related to a metric. The estimated values must be within the plausible space limited by the rules and facts. A better estimation is judged to be needed when any of the rules are not satisfied.

10.4.2 Proposed SDE Algorithm

This section presents a procedure for estimating any set of sparse, process data related to a given metric. However, appropriate techniques for analysis, collection and validation of data need to be applied for different metrics, *e.g.*, critical path method is used for task time analysis, *etc.*

Step 1: Collect information from a real process

Information collection is based on the metric analysis system. For instance, if the performance of the process is measured by the time accomplished for each process step; then, the metric 'time' is considered. Note that data collection and sparse data estimation can be performed at any time throughout the process execution.

Step 2: Check the necessity to perform SDE.

The next step is to determine the minimum required process steps that affect process performance with regard to a metric. There can be various approaches to identifying the minimum required process steps depending on the metric under consideration. For instance, CPM can be utilised to identify critical process steps when the metric under study is time. Structural importance analysis is also useful in identifying process steps.

Step 3: Populate the process model with measured data.

The process model is populated with the measured data by incorporating it into the process model.

Step 4: Perform initial estimation for missing data.

Once the first iteration of data collection for the identified set of process steps is completed, the next step is to determine whether the collected information is sufficient (complete) for overall process monitoring.

Step 5: Validate the initial estimation using the knowledge base.

If estimation of data is required due to missing data, constraints related to a specific metric that exist in a knowledge base are consulted in order to verify the validity of the set of process data. The information in the rule base can originate from a combination of the following:

- **expert knowledge** naturally exists in the mind of individuals familiar with the process, and it is gathered over a period of time by learning, trial and error, training, knowledge transfer, self-discovery, observation, *etc.*
- **network diagram and path generated rules.** Analysis of an IDEF3 model can result in the generation of a series of rules and facts about a process model which can be incorporated into a knowledge base. These rules and facts are in a general form and can be applied to any process.
- **statistics.** Statistical data, obtained by frequent repetition of a process, is a very reliable source for estimation. In such a situation the probability distribution of each metric for each process is identified. The distribution function, based on the context of the process and the metric being considered, can be one of the known distribution functions. For instance, task duration can be represented by a triangular distribution when it is known to be between two values, or in the presence of sufficient data collection; it can be represented by a probability distribution function.

Step 6: Develop better estimates if the estimates are not satisfactory

Despite the fact that the information collected from the actual process is not complete, there are various methods to estimate sparse information (Section 2.3.2). A set of algorithms is used to generate estimates for missing data; then, a simulation of the process is done to determine if the data set is consistent and satisfies all the necessary constraints.

Step 7: Complete model data

Once the estimates fulfil the knowledge base rules, the measured model data is complete and sent to the process comparison engine.

10.5 An Illustrative Example

In this section, the sparse data estimation method and the techniques explained in the previous section are illustrated through a sample process of design activity. The example provides an application of the aforementioned algorithm that can estimate task durations. The process model, as introduced by Larson and Kusiak (1996), is considered (Figure 10.1). Tasks 1 to 13 consist of the following design steps:

1-analyse product requirements *2-document internal design plan*
3- document external design plan *4-perform preliminary resource planning*
5-initiate type I design packet *6- initiate type II design packet*
7-develop hardware spec document *8- develop software spec document*
9-plan type II formal review meeting *10-develop spec summary document*
11- plan Type I formal review meeting *12-formal review with customer*
13- prepare and submit final design plan.

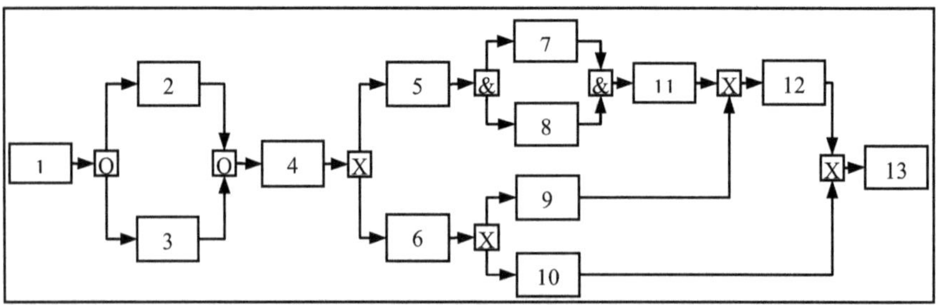

Figure 10.1. A sample IDEF3 process model with *and* (&), *or* (O) and *exclusive or* (X) junctions (Larson and Kusiak, 1996)

In the sample process model, the following considerations are assumed:

- Independency: Each activity in the process is independent from every other activity, which means that the start, finish, or execution of each activity has no impact on the start, finish, or execution of other activities.
- Non-Repetitiveness: In this example, it is assumed that activities are performed just once and there is no loop in the network.

- Abandon Resources: The resources to perform the tasks are assumed to be unlimited; therefore, there is no impact on the performance of the process due to the amount of resources.
- Smoothness: It is assumed that there is no specific control on each activity that would alter or modify its purpose.

For this example, task duration (time) is the metric.

Data collection

A first attempt is made to collect information from the real life process. As mentioned earlier, data collection and SDE can be performed at any moment during the execution of the process. A set of measurements is provided as a set of predetermined and normalised numbers with regard to duration. As can be seen in Table 2, the time expected to accomplish activities 2, 4, 8, and 9 are unknown.

Estimation Necessity Check

For the estimation necessity check, structural importance of the activities is considered. The structural importance value for an activity indicates its relative criticality with regard to alternative process paths. For instance, *performing the preliminary resource planning (task 4)* is mandatory for process success whereas *developing a software spec document (task 8)* could be omitted if *the initiate type II design packet (task 6)* is chosen instead of *initiate type I design packet (task 5).*

An algorithm has been developed for the calculation of structural importance for complex IDEF3 models. This algorithm first identifies all possible sets of activities (alternative paths) that ensure process attainability. A single binary logical statement describing the overall process attainability is then assembled. Every combination of the logical statement is considered to determine the structural importance values for each activity. Table 1 presents the structural importance values and the ranking for the activities of the sample process.

Table 10.1. Structural importance values for the sample process

Process	SIV	Rank		Process	SIV	Rank
1	255	1		8	15	7
2	85	4		9	45	6
3	85	4		10	141	3
4	255	1		11	15	7
5	15	7		12	63	5
6	231	2		13	255	1
7	15	7				

Process step 2 with a structural importance of 85 and process step 4 with a structural importance of 255 need their durations to be estimated. The estimation needs to be performed for process steps 8 and 9 as well; however, they are not structurally significant.

Initial Estimation

For the initial estimation, the real process model is populated with the measured data and initial estimations are made by considering expert opinion and/or the maximum expected duration of the project. The results are shown in Table 10.2.

Table 10.2. Measured and estimated durations for the sample process

Process	Measured	Estimated		Process	Measured	Estimated
1	5	4		8	?	5
2	?	7		9	?	2
3	4	3		10	4	2
4	?	5		11	7	9
5	5	5		12	3	1
6	3	4		13	2	3
7	2	3				

The estimated values are initial estimates that are required in order to simulate the process and verify that the estimates satisfy all constraints.

Validity Verification

The validity of the estimated values is verified vis-à-vis the knowledge base. In order to incorporate the time and structure related process constraints into the knowledge base, a set of facts and rules are generated utilising the facts and rules generation algorithms.

The facts generation algorithm creates the set of facts related to the sequential structure of the process model. The algorithm first calculates the critical path(s) (CP) of the process. The CP is calculated as:

$$CP = \{1, 2, 4, 5, 7, 11, 12, 13\}$$

Then, based on the calculated CP, a series of mathematical terms are generated by the algorithm. These terms denote the start and finish times of each activity with regard to duration as well as the sequential relationships between activities. 29 facts are generated for the sample process. Examples include:

$$FT[1] = ST[1] + D[1]; \; (finish[1] =]; start[1] + duration\ [1])$$

$$ST[4] >= min(FT[2], FT[3]$$

$$ST[4] <= max(FT[2], FT[3])$$

The rules generation algorithm creates structure related process rules. These rules denote the occurrence of each activity related to the occurrence of alternative paths and other activities as a series of if-then-else statements. In a design process, for instance, if the *document external design plan* (task 3) does not occur, then, the *analyse product requirements* (task 1) does not occur. Examples within the 131 facts generated for the sample process include:

If process step 3 occurs, then process step 1 occurs;

If process step 2 does not occur, then path 6 does not occur.

These rules and facts along with the available expert and statistical information in the knowledge base define a plausible space that delimits the duration values for each task. The knowledge base now incorporates all available data for the validity check and advanced estimation of the process information.

For the purpose of this study, FuzzyCLIPS software was used in order to verify the validity of the estimated values. According to the knowledge base, the estimation for process step 2 is not valid and a new estimation is required.

Better Estimation Development

In order to obtain better estimates, fuzzy reasoning techniques are applied. As mentioned earlier, expert knowledge base software such as FuzzyCLIPS, allows representing and manipulating fuzzy facts and rules generated in the SDE process. The knowledge base is used here to incorporate different rules and information associated with the process steps (tasks or fragments of tasks). Fuzzy facts, *e.g.*, 'low' or 'high' durations, are considered along with the steps of the process. Hence, the process fragment including task 2 and task 4 has fuzzy information associated with it. The defuzzier built within FuzzyCLIPS is used to extract crisp values of these durations. For instance, the duration of process step 2 is estimated as 5.4. Since this new estimation satisfies all the constraints in the knowledge base, the estimation for the sample model is complete.

The measured process is completed with estimated values; thus, the measured and ideal process models have identical structures and complete datasets. Process comparison can begin.

10.6 Conclusion

This paper focused on sparse data estimation as the main component of an IPMS to cope with the chronic problem of estimating missing data. The method uses process structure and rules to estimate and re-estimate task metrics as a knowledge process unfolds.

Several procedures in the SDE method were introduced in order to automatically generate rules from a network diagram as well as from process metrics and subject matter experts to obtain a rule-base. Any estimation that satisfied the constraints in the constructed rule-base was considered as a good estimate. Artificial intelligence and fuzzy reasoning methods were utilised in order to alter non-satisfactory estimates so that they would be acceptable.

The proposed SDE method allows early estimation of the progress of design processes as opposed to traditional project monitoring methods where a complete set of information is identified towards the end of project completion. Furthermore, the automatic rule generation techniques within the SDE method can lead to the continuous process monitoring from early stages of execution.

Although the illustrative example demonstrates the applicability of the SDE method for the metric 'time', the proposed method is generic and can be applied using different design metrics, such as process quality or product performance. Note that adapted techniques for rule generation and validity estimation need to be determined for different metrics.

Although the case-example provides verification that the Sparse Data Estimation method and the proposed algorithms can estimate task durations, the method is designed to use any metric; however, this needs to be tested in real cases with multiple metrics for a complete demonstration. Indeed, there is a need to demonstrate the method on real knowledge intensive processes underlining how the estimation and validation of the metrics are undertaken relying on the knowledge based and the actual process measurements. Also, the application of the SDE approach considering quality, reliability and performance metrics is paramount for the full demonstration of the approach. These tests should also reveal the cost for building and operating a multi-metric sparse data estimation model.

The SDE method was created to assist monitoring of knowledge work processes since process parameters are not easily measured, and thus, there is usually missing process data. In this paper, the method was demonstrated for design processes. However, the SDE method can be used for any system which can be represented as a network. Network rules along with rules for the metric of interest can be used as constraints to estimate missing data.

Future work will develop an extension to the structural importance algorithm to deal with cyclic processes where feedback and iterations are ubiquitous, such as design. The method also needs to be tested for its limit on sparseness, *i.e.*, how much data can be missing and are there any process features where data cannot be missing.

10.7 References

Dempster AP, Laird NM, Rubin DB (1977) Maximum likelihood for incomplete data via the EM algorithm. Journal of the Royal Statistical Society, B 39: 1–38

Hebert JE (1975) Critical path analysis and a simulation program for resource-constrained activity scheduling in GERT project networks. Unpublished PhD Dissertation, School of Industrial Engineering, Purdue University, IN, US

Karsai G, Sztipanovatis J, Padalkar S, Bigel C, Miyaska N, Okuda K (1992) Model-based intelligent process control for cogenerator plants. Journal of Parallel Processing and Distributed Computing, 15: 90-102

Lari K (2004) Sparse data estimation for knowledge processes. PhD Thesis, McGill University, Montreal, Canada

Larson N, Kusiak A (1996) Managing design processes: A risk assessment approach. IEEE Transactions on Systems, Man and Cybernetics - Part A - Systems and Humans, 26(6): 749-760

Lenat DB, Guha R (1990) Building large knowledge bases. Addison Wesley, MA, US

Little RJA, Rubin DB (1987) Statistical analysis with missing data. John Wiley and Sons Inc, Hoboken, NJ, US

Malcolm DG, Roseboom JH, Clark CE, Fazar W (1959) Application of a technique for research and development program evaluation. Operations Research, 11(5): 646–669

Mayer RJ, Cullinane TP, deWitte PS, Knappenberger WB, Perakath B *et al.* (1992) IDEF3 process description capture method report. Knowledge Based Systems, Incorporated, TX, US

Nayestani N (2002) A process comparison algorithm. Masters Thesis, McGill University, Montreal, Canada

Ross DT, Schoman KE (1977) Structured analysis for requirements definition. IEEE Transactions on Software Engineering, SE-3 6(15)

Tanner MA, Wong WH (1987) The calculation of posteriori distributions by data augmentation with discussion. The Journal of the American Statistician Association, 82: 528–550

Turksen IB, Tian Y (1995) Two-level tree search in fuzzy expert systems. IEEE Transactions on Systems, Man and Cybernetics, 25(4): 555-569

Turksen IB, Zhao H (1993) An equivalence between inductive learning and pseudo-boolean logic simplification: A rule generation. IEEE Transactions on Systems, Man and Cybernetics, 23(3): 907-917

Wynn DC (2007) Model-based approaches to support process improvement in complex product development. PhD Thesis, Cambridge Engineering Design Centre, University of Cambridge, Cambridge, UK

Chapter 11

Deriving Event Graphs through Process Mining for Runtime Change Management

H. Zhang, Y. Liu, C. Li and R. Jiao

11.1 Introduction

Engineering Changes (ECs) refer to any modifications in forms, fits, functions, materials, services and so on in product design, development and the subsequent manufacturing operations. In a broad sense, EC is an alteration made to parts, drawings or softwares that have already been released during the product design process. The change can be of any size or type. The change process can involve any number of people and can take any length of time (Jarratt, 2004). Business changes, organisational changes, schedule changes, order changes and so on are not directly related to product design and therefore are not usually considered as ECs. However, they may be consequences of ECs or they may become the dominating causes for ECs. In practice, any ECs occurring after the design which are approved and released are referred to as late changes. On the contrary, any ECs triggered before the design is approved and released are named early changes. Late ECs are much more costly and time-consuming to implement than early changes. The best approach in managing ECs is to avoid them as far as possible. However, ECs are often deemed indispensable in many real scenarios. Thus, it has become a research issue as to how ECs can be managed effectively and efficiently.

Engineering Change Management (ECM) is of major concern to all companies that design and manufacture goods. This message has been repeatedly highlighted by a number of comprehensive surveys and in-depth case studies (Mall *et al.*, 1992; Huang and Mak, 1997, 1998; Huang, 2003). Managing ECs may cost as much as up to 10% of annual turnover and designers may spend up to 40% of their prime time in making ECs (Mall *et al.*, 1992). There have also been a number of industrial case studies on the ECM subject. Stories are similar to survey results. In summary, these industrial studies and case studies have highlighted a clear message that EC is a serious issue that cannot be under-estimated and ECM is of major concern to most companies that design and manufacture products.

Characterised by geographical, temporal, functional and semantic distribution, today's enterprise models engage multiple design teams with heterogeneous skills, cooperating together in order to achieve a global optima in design (Chira *et al.*, 2005). In such a context, the management of engineering change becomes more complicated. Today's ECM is regarded as an integrated and crucial part of the product design and development process, instead of being considered separately. The presentation of change in various product development projects differs significantly, making it difficult to develop unique processes and tools to deal with them. Due to the importance, complexity and current practice of ECM, industry needs a better ECM. However, existing approaches in handling ECs possess insufficiency in providing a thorough understanding of EC system modelling, its dynamic nature, its reasoning and change propagation. As a result, it severely limits the application of ECM in industry practice. In this paper, we propose a framework for ECM technique based on event graph (Schruben, 1983). Our basic idea is to derive event graphs from the workflow logs through process mining (also known as workflow mining) (Agrawal *et al.*, 1998). On the basis of the event graph model, we intend to carry out further studies about ECM. For example, we analyse the cost, schedule and risk of the product development processes. Another aspect is to track and analyse different processes for change prediction, change distribution, change propagation and change impact and so on. All these researches will help manage ECs more effectively and efficiently.

The rest of this paper is organised as follows. Section 11.2 reviews existing studies on ECM. The proposed approach is elaborated in Section 11.3. A preliminary study of managing ECs based on event graph is illustrated in Section 11.4. Conclusions and future work are given in Section 11.5.

11.2 Related Work

In terms of ECM modelling, which captures the process, entities and solutions involved in dealing with engineering changes, existing approaches can be classified into three main streams:

- Process based model - it provides a good top-level view of the design process with good visibility of different design goals. The process and time-dependent nature of a design can be modelled by activities or task networks and associated algebraic representations. These representations rely on the relationships and connections between various activities (Eppinger *et al.*, 1994; Clarkson and Hamilton, 2000; Earl *et al.*, 2001). During the last couple of decades, many different formalisms have been raised to represent the process based model. State transition diagrams of the object-oriented analysis method were presented (Mellor and Shlaer, 1992). A Prism model of changes was proposed in software systems supported by two environment infrastructures called the dependency structure and the change structure (Madhavji, 1992). A change propagation model was established during software maintenance and evolution (Rajlich, 1997). A UML based workflow integrated system was developed for change management including functions like complex product data management, versioning, configuration management, change notification and so on (Ivins *et al.*, 2000). A data model was built to track design changes (Xie, 2001). Recently, an Integer Linear Programming (ILP) approach was introduced for design change (Farinaz *et al.*, 2002). Jarratt, Clarkson and Eckert discussed some of the different engineering change processes proposed in literature and outlined a generic process (Jarratt *et al.*, 2005).

- Task based model - it supports multi-parameter problems. However, it is generally inflexible and project specific. Popular task-based models include Design Structure Matrix (DSM), Signposting and Petri nets. DSM was widely used to analyse tasks and change propagation (Eppinger *et al.*, 1994; Ulrich and Eppinger, 2000). A model called Signposting was developed for the engineering design context (Clarkson and Hamilton, 2000; Stacey *et al.*, 2000). A methodology for modelling and simulating product development process was proposed based on the extended stochastic high-level evaluation Petri nets (Yan *et al.*, 2003).
- Parameter based model - it can be regarded as a collection of design tasks characterised by their required input parameters. A parameter is "an aspect of a design that needs to be determined, and hence embodies a decision about the design" (Stacey *et al.*, 2000). A methodology called Change Favourable Representation (C-FAR) was proposed using existing product data information to capture possible change consequences to a product. It adopted the existing product information model to facilitate change representation, propagation and qualitative evaluation (Cohen *et al.*, 2000). Li and Offutt proposed a data dependency graph to analyse change impact for object-oriented software (Li and Offutt, 1996).

Besides such researches about ECM, some ideas and technologies of product development are also related and helpful for managing ECs. Ottosson presented Dynamic Product Development (DPD) with the management principle "Management by Walking Around" so as to get unfurbished and extensive feedback in real time with the aim of being able to make adjustments without delays (Ottosson, 2004).

Along with the theoretical research, several ECM systems have been developed, such as a change management framework based on object-oriented database system (Huh and Rosenberg, 1996), a web-based system for ECM (Huang *et al.*, 2001), PDM/MRP integration framework to evaluate the influence of engineering change on inventory scrap cost (Ou-Yang and Cheng, 2003).

Although many existing studies have made significant advancements in EC modelling and its system development, using the process-oriented approach: the primary means in current ECM, a notable difficulty is its insufficient ability in providing a thorough understanding of EC system modelling, its dynamic nature, reasoning and propagation. Hence, it severely limits the application of EC management in industry practice.

11.3 An ECM Framework Based on Event Graph

In order to manage ECs better, the focus of our research is to propose a dynamic model using an event graph to enhance the efficiency and quality of ECM.

The proposed event graph based ECM framework is shown in Figure 11.1. Any information system using transactional systems such as enterprise resource planning, customer relationship management or workflow management systems will provide workflow information in some form and to some extent, such as tasks available, their events and the details of these events like starting/ending timestamp. A workflow log is often a consistent XML document containing such information (van der Aalst *et al.*, 2003). In our study, we aim to extract information from such logs and derive an event graph through process mining. Based on the event graph model discovered, further works of ECM are proposed like process tracking and change impact analysis. In the end, the analytical results can be used for process reengineering in order to optimise enterprise processes.

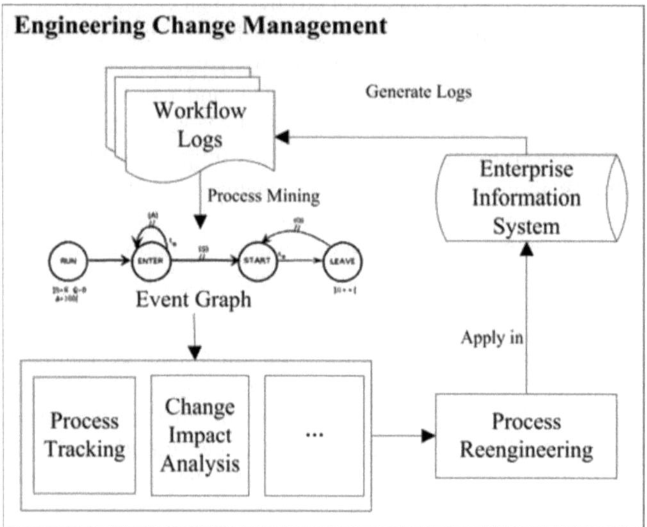

Figure 11.1. Framework for ECM based on event graph

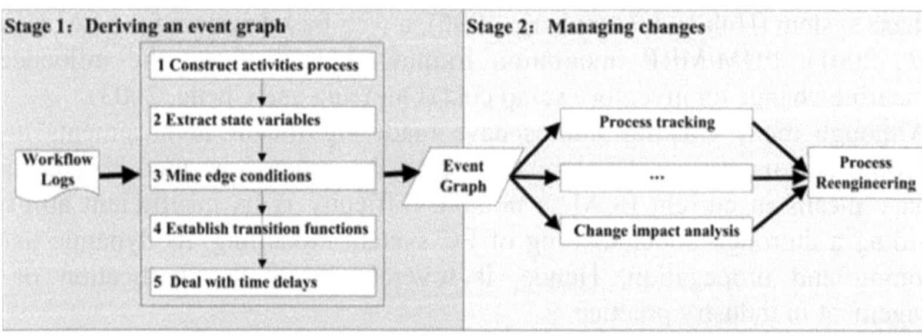

Figure 11.2. Two stages for the proposed framework

According to the generic nature of functions and techniques adopted in our study, our proposed framework can be separated into two main stages shown in Figure 11.2, Stage 1: deriving an event graph through process mining, and Stage 2: managing changes based on the event graph generated. The steps of event graph derivation through process mining are elaborated in Section 11.3.1.

11.3.1 Deriving an Event Graph through Process Mining

In our research, we view business process as a combination of different activities with a structure describing their logical order and dependency. In this stage, we aim to model a process by constructing an event graph where the graph itself is discovered through process mining. The event graph formed will help us to gain more insights of the process under investigation and prepare for its further analysis.

11.3.1.1 Introduction to Event Graph

Event graph is first presented by Schruben to develop simulations of discrete event systems (Schruben, 1983). The system dynamics are characterised by events that change the state of the system and the logical and temporal relationships among events. Figure 11.3 shows the fundamental construct of event graph and is interpreted as follows: the occurrence of event A causes event B to be scheduled after a time delay t, providing condition S>0 is true (after the state transitions for event A has been made, S is a state variable). Later in 1988, Schruben presented simulation graph as the mathematical formalisation and extension of event graphs. A simulation graph is an ordered quadruple, where the four elements in the quadruple represent the sets of vertex, scheduling edges, cancelling edges and incidence functions respectively (Schruben and Yucesan, 1988). Based on the simulation graph, a simulation model can be defined with the following key elements, which also play an important role in our model.

- Event vertex: In simulation graphs, every vertex stands for an event and it must be regarded as instantaneous. So events in workflow logs are mapped into event vertexes.
- State variable: States of a process are reflected by state variables. They can be altered by the event vertexes and used in edge conditions. In our study, we use them to integrate enterprise information.
- Edge condition: It is relatively easy to build the activity processes because edge conditions are used to control the process flow.
- Transition function: Changes of state variables made at event vertexes are referred to as transition functions.
- Time delay: They can be specified at edges and they form an indispensible part of a global simulation clock.

After Schruben and Yucesan's proposal of simulation graph, researchers in the community actually did not make any distinction between these two graphs. Case studies and simulations were reported (Buss, 1995; Arnold, 1996).

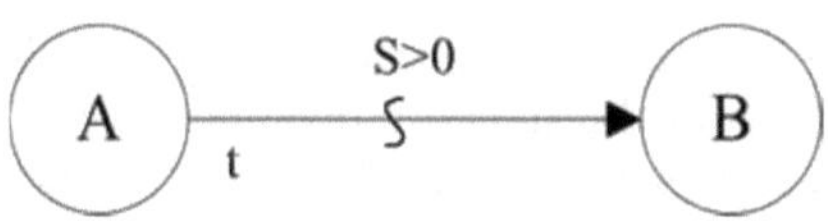

Figure 11.3. Fundamental event graph construct

11.3.1.2 Merits of Event Graph

Event graph is a combination of graph-based formalism and rule-based formalism (Lu and Sadiq, 2007), with its visual appeal of being intuitive and explicit for graph-based formalism and a good presentation for rule-based formalism. Event graph models have been demonstrated to be able to model any system that can be implemented on a modern computer (Eric *et al.*, 2005). Here we summarise the merits of event graph compared to Petri nets, PERT and Critical Path Method (CPM). Firstly, event graphs have powerful expression as they pull together the various aspects of enterprise activities, with their different elements, in a concise manner so that it is able to track and analyse the process performance. However, most Petri nets are only concerned with the activity schedules for enterprise processes. PERT has no consideration of resource so that it cannot solve problems of resource competition and activity conflict. Secondly, event graph possesses edge conditions to deal with the logical relationships between activities, so that only the causality relations needs to be considered in constructing activity processes from workflow logs. But in Petri net, it is not easy to discover logical relations such as parallelism and choice. It is even unpredictable for CPM. Thirdly, it is flexible to arrange the control flow with the help of edge conditions in event graph. But it is quite an effort to establish complex logical relations between activities in both PERT and CPM.

11.3.1.3 Steps for Constructing Event Graph from Logs

Besides the activity process, in addition, event graphs combine other enterprise information so as to describe the process more precisely and comprehensively. In this case, the approach for constructing event graph through process mining consists of several steps for different components. Here we summarise the five main steps for our approach shown in Figure 11.2.

(1) Construct activity process. Unlike the Petri net, event graph does not need to concern itself with the different types of logical relations such as choice and parallelism. We only need to construct the causality relations, because the flow control relies on the edge conditions. Therefore, activity processes with causality relations form the crucial architecture of an event graph. One issue that we have paid attention to is that the event vertexes are regarded as instantaneous. When we need to consider the duration of an event, it is better for us to set two vertexes with a starting vertex and an ending vertex.

(2) Extract state variables. The state variables are crucial in event graph. They are usually mapped from the enterprise information in the logs. In this way, it is able to combine both data structure and behaviour in a single entity. In this step, we need to extract all the information in the workflow logs and determine how they could be mapped into state variables. After getting raw data from the logs, we first carry out in-depth statistic analysis with the help of tools like Weka (Holmes *et al.*, 1994). Based on the statistical results, we determine which data features should be mapped into the state variables and what their types and ranges are.

(3) Mine edge conditions. Edge condition is an important component of event graph, because it plays an important role in controlling the process flows so that we need not pay much attention to the logical relations between activities. Edge conditions should be considered when an event has two or more subsequent events.

Thus, we extract all these events, their subsequent events and all the interrelated state variables. The state variables are kept as attributes and the names of subsequent events are kept as class labels. By this means, all the information is extracted and transformed into a standard .arff file as training data for classification in Weka by which a set of classification rules are generated. We then take the rules generated as edge conditions. Such rules can surely be verified by human experts later on.

(4) Establish transition functions. Transition functions represent how an event vertex affects state variables. However, it is not so easy to mine transition functions automatically. This is actually because enterprise information is complex due to numerous names, different types, complex relations and so on. We consider two situations. The first is when a state variable shows up for the first time at an event vertex. In this case, we consider that the event vertex will alter the state variable from its default value (for example, zero for numeric) to a random value but still in the corresponding range. The second is even though the state variables have not appeared for the first time but are changed. In this case, we have made a simple assumption that all the state variables, before and after a certain event, are basically located in the same semantic layer and are in the same order so that the transition function can be realised through some simple mathematic formation like addition and subtraction. Therefore, we can deal with the transition functions for the latter situation with multi-linear fitting, where the coefficients in the multi-linear equations are 0, 1 or -1. For example, the state variable A changes after an event e_i. There are other states variables B, C and D before the occurrence of e_i. So the problem is to find the fitting coefficients for the Equation 1.1:

$$A_1 = a \cdot A_0 + b \cdot B + c \cdot C + d \cdot D \tag{1.1}$$

Where A_0 and A_1 are the values of state variable A before and after the occurrence of e_i respectively. Coefficients a, b, c and d are any of (0, 1 or -1).

(5) Deal with time delays. Temporal information is also a significant factor for business process. However, it has not been made full use of in the majority of existing studies. Event graph contains time delays on edges and through which it obtains a global clock in simulation. It is also useful for other in-depth studies, *e.g.* mining important information from event graph, predicting engineer change impact and so on. In event graph, every event vertex is regarded as instantaneous and time delays are expressed on edges. We have thought about two ways to deal with the delays. One is that the time delay between event i and event j denotes for the interval from the start of event i to the start of event j. However, if there is no time gap, it can also be considered as the duration of event i. The other way is to map an event into two vertexes each with a starting vertex and an ending vertex. In this case, the time delay between them is taken as the duration of the event. Although this is complex, it is more precise than the previous. This is a more complex but more exact way.

After the five steps above, we obtained the activity process, state variables, edge conditions, transition functions and time delays. Thus, we are able to visualise the graph in Sigma, a modelling and simulation tool for event graph (Schruben, 1995).

11.4 Managing Changes Based on Event Graph

With the constructed event graph model, we can advance to managing changes based on it. At this stage we aim to research and investigate EC problems from the modelling perspective, a complete different angle. One important problem in change management is change propagation. Change propagation is "the process of forwarding a data change from a source system to all dependent systems" (Rantzau *et al.*, 2002). Different entities (including requirements, specifications, BOM, product parts, components and assemblies, *etc.*) in the product development process are dependent upon each other. Changes made to any of these will result in alterations on other parties. Change propagation is a long, difficult, and error-prone incremental change activity (Rajlich and Gosavi, 2004). Owing to its importance and complexity, we attempt to carry out the impact analysis in change propagation. Change impact analysis, as discussed here, refers to the analysis of how ECs affect other entities in the process like product development. Based on the change impact analysis, we can modify the schedules and arrangements for process re-engineering. Thus, it is able to find the most efficient and appropriate working procedure, even for the whole enterprise process.

To briefly illustrate our idea, Figure 11.4 shows a constructed process model with event graph drawn in Sigma. Because the event vertexes are regarded as instantaneous, there are two vertexes for an event named at the beginning with "StartX" and ending "EndX". They stand for the start point and the end point for an event. By this means, the time delay of an edge between a start point and an end point can be considered as the duration of the event.

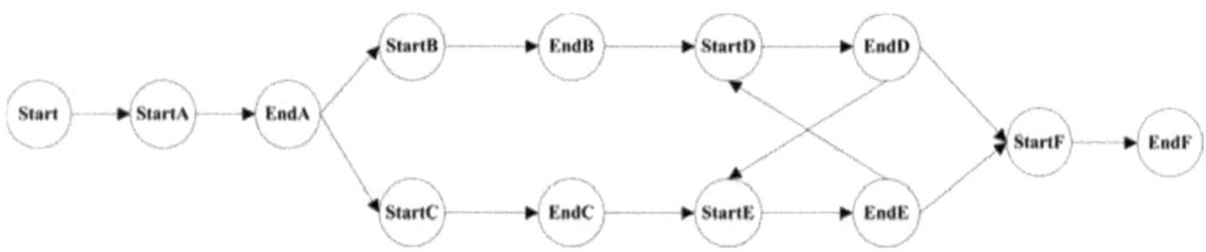

Figure 11.4. An example of event graph model in Sigma, the first vertex named "Start" is the starting point for event graph in Sigma without real meaning

This model is intended to have two schedules for this process shown in Figure 11.5. However, the schedules can not be directly seen from the graph. This is because the state variables and edge conditions, which are used to control the flow, are not directly shown in the main window of Sigma but are actually embedded in other windows behind.

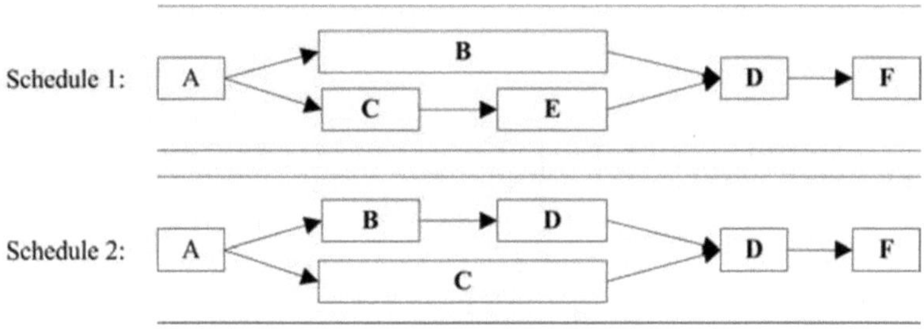

Figure 11.5. Two schedules for the model

For better visualisation, we draw another full picture for this model in Figure 11.6, where VALUE1, VALUE2 and VALUE3 are the state variables of said flow control. The notations near the vertexes are transition functions. The notations with RND along the edges are time delays. RND is a Sigma function that offers a random number between 0 and 1. For example, 10+5*RND denotes for a random number anywhere between 10 and 15. Here we only focus on the durations of events and assume no time gaps between events. Under this assumption we place time delays between the starting and ending point of an event.

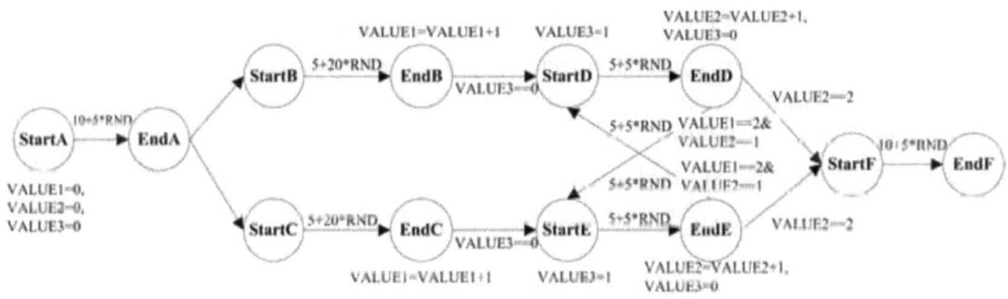

Figure 11.6. The full graph of the event graph model

Now a change occurs because event B is going to take more time due to the interruption of resource supply. Initially, it needs 5 to 25 time units. Currently, it should be extended to 25 and 40 units. In order to understand the impact of this change, we simulate the processes of both schedules before and after the change is made in Sigma and observe the change of process durations under each situation. Figure 11.7 shows the comparison.

From Figure 11.7, before the change, we can hardly see the duration differences between these two schedules. It is obvious that both durations become longer if event B takes more time for execution. However, the ranges of duration increase are obviously distinct. It is easy to see that the increase of Schedule 2 is significantly higher than the one of Schedule 1.

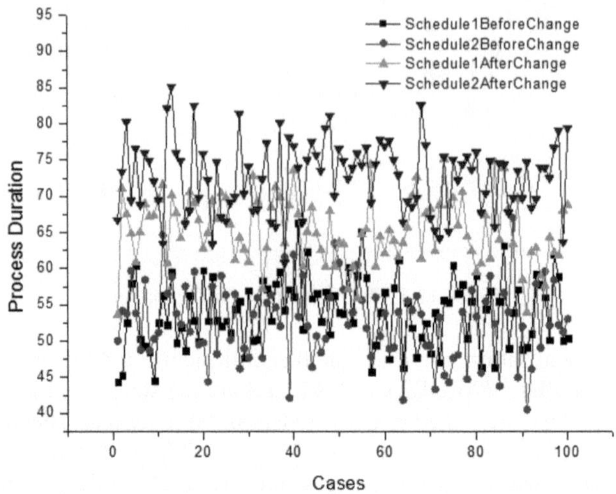

Figure 11.7. Comparison of process durations for two schedules before and after the change

In order to make the result comparison more comprehensive, we list all the minimum, maximum and average time before and after the changes take place in each schedule in Table 11.1. Obviously the average process durations for these two schedules are very close to each other. However, the durations of both schedules are increasing after the change was implemented. Meanwhile, it takes more time for schedule 2 than schedule 1. Through this comparison, we can conclude that it sounds better to choose schedule 1 for the process after the change, because it can make the process more efficiently and save time for execution.

Table 11.1. Time information for two schedules before and after the change

	Schedules	**Min Time**	**Max Time**	**Mean Time**
Before change	Schedule1	44.34	66.37	53.970
	Schedule2	40.52	66.99	52.738
After change	Schedule1	53.63	75.13	65.407
	Schedule2	60.74	85.13	72.725

11.5 Conclusions and Future Work

A flexible, adaptable and dynamic model is crucial for supporting ECM in industry. In this paper, we have proposed a framework for deriving event graphs through process mining for runtime change management. Detailed analysis and comparison are given to show the modelling capability of event graph for ECM. In particular, such a building approach has integrated all the enterprise process related information in a concise model. Further works, such as change propagation and change impact analysis, on ECM are briefly presented. We have also illustrated how we can manage ECs based on event graph. Lastly, we believe that efforts committed in ensuring a formal and sound evaluation of our work should be constantly pursued. In the future, we shall focus more on the evaluation and validation of the proposed event graph modelling approach using a large volume of real-world data.

11.6 Acknowledgement

The work described in this paper was supported by a research grant from the Hong Kong Polytechnic University, Hong Kong SAR, China (Grant No: A-PD0M).

11.7 References

Agrawal R, Gunopulos D, Leymann F (1998) Mining process models from workflow logs. In: Schek HJ (ed.) Proceedings of the 6th International Conference on Extending Database Technology: Advances in Database Technology, Springer Verlag, Heidelberg, Germany

Arnold HB (1996) Modeling with event graphs. In: Proceedings of the 28th Conference on Winter Simulation, Coronado, CA, US

Buss AH (1995) A tutorial on discrete-event modeling with simulation graphs. In: Proceedings of the 27th Conference on Winter Simulation, Arlington, VA, US

Chira C, Chira O, Roche T (2005) Multi-agent support for distributed engineering design. Springer Berlin/Heidelberg, Berlin, Germany

Clarkson PJ, Hamilton JR (2000) 'Signposting', a parameter-driven task-based model of the design process. In: Research in Engineering Design 12(1): 18-38

Cohen T, Navathe SB, Fulton RE (2000) C-FAR, change favorable representation. Computer Aided Design 32: 321-338

Earl CF, Eckert CM, Johnson JH (2001) Complexity of planning in design. In: Proceedings of the 2001 ASME Design Engineering Technical Conferences (DETC'01), Pittsburgh, PA, US

Eppinger SD, Whitney DE, Smith RP, Gebala D (1994) A model-based method for organizing tasks in product development. Research in Engineering Design 6(1): 1-13

Eric LS, Lee WS, Enver Y (2005) On the generality of event-graph models. INFORMS Journal on Computing 17(1): 3-9

Farinaz K, Jennifer LW, Jessica F, Miodrag P (2002) ILP-based engineering change. In: Proceedings of the 39th Annual Design Automation Conference, New Orleans, LA, US

Holmes G, Donkin A, Witten IH (1994) WEKA: a machine learning workbench. In: Proceedings of the 1994 2nd Australian and New Zealand Conference on Intelligent Information Systems, Brisbane, Australia

Huang GQ (2003) Current practice of engineering change management in Hong Kong manufacturing industries. Journal of Materials Processing Technology 139: 481-487

Huang GQ, Mak KL (1997) Computer aids for engineering change management. In: Proceedings of International Conference on Computer-Aided Production Engineering, Warsaw, Poland

Huang GQ, Mak KL (1998) Computer aids for engineering change control. International Journal of Materials Processing Technology 76: 187-191

Huang GQ, Yee WY, Mak KL (2001) Development of a web-based system for engineering change management. Robotics and Computer Integrated Manufacturing 17: 255-267

Huh S, Rosenberg DA (1996) A change management framework: Dependency maintenance and change notification. Journal of Systems and Software 34: 231-246

Ivins WK, Gray WA, Miles JC (2000) A process-based approach to managing changes in a system to support engineering product design. In: Topping BHV (ed.), Developments in Engineering Computational Technology. Civil-Comp Press, Edinburgh, UK

Jarratt TAW, Clarkson PJ, Eckert CM (2005) Engineering change. In: Clarkson PJ, Eckert CM (eds.) Design process improvement - a review of current practice. Springer Verlag, London, UK, pp 262-285

Jarratt TAW (2004) A model-based approach to support the management of engineering change. PhD Thesis, Cambridge Engineering Design Centre, University of Cambridge, Cambridge, UK

Li L, Offutt AJ (1996) Algorithmic analysis of the impact of changes to object-oriented software. In: Proceedings of the IEEE International Conference on Software Maintenance (ICSM'96), Monterey, CA, US

Lu R, Sadiq S (2007) A survey of comparative business process modeling approaches. In: Proceedings of the 10th International Conference on Business Information Systems (BIS 2007), Posnan, Poland

Madhavji NH (1992) The Prism model of changes. IEEE Transactions on Software Engineering 18: 380-392

Mall R, Hughes D, Shlaer H (1992) The role of the bill-of-materials as a CAD/CAPM interface and the key importance of engineering change control. Computing & Control Engineering Journal 3(2): 63-70

Mellor SJ, Shlaer S (1992) Object lifecycles: Modeling the world in State. Prentice Hall, NJ, US

Ottosson S (2004) Dynamic product development - DPD. Technovation 24(3): 207-217

Ou-Yang C, Cheng MC (2003) Developing a PDM/MRP integration framework to evaluate the influence of engineering change on inventory scrap cost. The International Journal of Advanced Manufacturing Technology 22: 161-174

Rajlich V (1997) A model for change propagation based on graph rewriting. In: Proceedings of the IEEE International Conference on Software Maintenance (ICSM'97), Bari, Italy

Rajlich V, Gosavi P (2004) Incremental change in object-oriented programming. IEEE Software 21(4): 62-69

Rantzau R, Constantinescu C, Heinkel U, Meinecke H (2002) Champagne: Data change propagation for heterogeneous information systems. In: Proceedings of the 28th International Conference on very Large Data Bases, Hong Kong, China

Schruben LW (1983) Simulation modeling with event graphs. Journal of Communications of ACM 26(11): 957-963

Schruben LW (1995) Graphical simulation modeling and analysis: using SIGMA for windows. Boyd & Fraser, Danvers, MA, US

Schruben LW, Yucesan E (1988) Simulation graphs. In: Proceedings of the 20th Winter Simulation Conference, San Diego, CA, US

Stacey MK, Clarkson PJ, Eckert CM (2000) Signposting: An AI approach for supporting human decision making in design. In: Proceedings of the ASME International Design Engineering Technical Conference (IDETC/CIE2000), Baltimore, MD, US

Ulrich KT, Eppinger SD (2000) Product design and development, 2nd edn. McGraw-Hill, NY, US

van der Aalst WMP, van Dongen BF, Herbst J, Maruster L, Schimm G, Weijters AJMM (2003) Workflow mining: a survey of issues and approaches. Journal of Data & Knowledge Engineering 47(2): 237-267

Xie H (2001) Tracking of design changes for collaborative product development. In: Proceedings of the 6th International Conference on Computer Supported Cooperative Work in Design, London, Ontario, Canada

Yan HS, Wang Z, Jiao XC (2003) Modeling, scheduling and simulation of product development process by extended stochastic high-level evaluation Petri nets. Robotics and Computer-Integrated Manufacturing 19(4): 329-342

Chapter 12

Assessment of Design Tools for Knowledge Capture and Re-use

A.V. Gokula Vijaykumar and A. Chakrabarti

12.1 Introduction

Designers primarily use paper and pencil or software packages for articulating their ideas. As the need for capture and re-use of the knowledge generated during designing becomes stronger, this propels the need to develop more sophisticated mechanisms for supporting these knowledge processes. We argue that interactions occurring during the design process play a vital role in knowledge processing. We define interaction as a mutual or reciprocal action or influence of agents (perceptible objects through which designing occurs). Since interactions especially between designers and tools is likely to have a substantial impact on the knowledge processes, it is necessary to change or modify existing tools, or develop new tools, to increase the effectiveness of capturing and reusing knowledge.

Review of literature on tools for capture and re-use of knowledge reveal that a variety of new tools has been developed in the last few years (Mugellesi-Dow *et al.*, 2008). However, while the importance of knowledge management (KM) tools has been stressed in organisations, the implications of usage of these tools on KM performance are yet to be studied in detail. Even though many tools have been implemented in organisations, the full potential of the tools could be extracted only when these are applied correctly and integrated into the day-to-day processes of designers. Many software tools have been studied and reported in literature, such as document management systems, decision support systems, groupware, content management systems, and electronic directories. However, the impact of design equipment such as paper, digital tablets, or Personal Digital Assistants (PDAs), on knowledge processes has not been studied in detail. Most tools have been evaluated in terms of their usability. Pinelle and Gutwin (2000) summarise the evaluation factors considered in the various studies as:

- organisational impact/impact on work practices;
- end product produced through using the software;
- efficiency of task performance using software;

- user satisfaction with the software;
- task support provided by the software;
- specific features of the software interface,
- patterns of system use;
- user interaction while using the software.

Cugini *et al.* (1999) propose a frame divided into four levels for evaluating tools: requirements, capability, service and technology. Based on a review of literature, Taylor (1996) identifies six broad categories of criteria that users employ to select information systems: ease of use, noise reduction, quality, adaptability, time-saving, and cost-saving. Vizcaíno *et al,* (2005) argue that while some of these criteria are related to KM, none is specifically focused on evaluating information systems from a KM perspective. In this paper we intend to evaluate tools from the perspectives of knowledge capture and re-use. Since interaction is a main focus of this work, influence of tools on knowledge capture and re-use during interactions is studied. This study, it is hoped, will help moderate the current emphasis of KM systems on the technical stages rather than on addressing the needs of designers. Duffy and Duffy (1996) propose Engineering Design Re-use Process Model to effectively use learned knowledge. But influence of interactions is not studied.

The work contains two consecutive studies. In the first study, seven individual tools (with which two designers worked individually in conceptual design sessions) have been tested for their effectiveness on knowledge processes during designing. The tools used were Paper, Mobile e-Notes Taker™, Tablet without viewing facility, Wacom™ Tablet with viewing facility, Computer with Rhinoceros™ CAD package, Mobilis™ personnel digital assistant (PDA), and Sony Ericsson P1i™ PDA (Encore, 2008; Hi-Tech Solutions, 2008; iBall, 2008; Rhinoceros, 2008; Sony Ericsson, 2008; Wacom, 2008). The specifications of the tools are given in (Ideas Lab - Equipment, 2010). Based on this evaluation, the three, top rated individual tools were selected for a second, more in-depth study. The rest of the paper elaborates the research questions, methodology used, the results obtained from the two studies and the conclusions drawn.

12.2 First study: Research Questions and Methodology

The primary focus of this paper is on studying the influence of the following tools on knowledge processes: Paper, Mobile e-Notes Taker™, Digital Tablet without viewing facility, Digital Tablet with viewing facility, Computer with Rhinoceros™ CAD package, Mobilis™ Personnel Digital Assistant (PDA) and Sony Ericsson P1i™ PDA. Selection of these tools has been influenced by their potential to replace pencil and paper which are currently the most commonly used aid for conceptual design. Even though support is required for all design stages, we took the task clarification and conceptual stages as the first stages to be considered in initiating an assessment of these tools, because of the impact these stages have on the design while being the least time consuming to carry out.

Since detailed evaluation of all seven tools is time consuming, the first study is conducted to help identify the three best tools for detailed evaluation. The first study involves assessing designers' interactions with these tools for the following knowledge processes: generation, modification/editing, capture, and re-use. These activities are defined as follows. Knowledge generation includes all activities of designers that aid in solving the design problem; knowledge capture includes the activities of designers that aid to record the articulated knowledge through the support tool; knowledge modification includes the activities of designers which aid to change the recorded knowledge; knowledge re-use includes the activities of designers that aid the use of the recorded knowledge for different design tasks or problems. Re-use includes reading, analysing the material and selecting some pieces of valuable material. The main focus of this work is on knowledge capture and re-use, while generation and modification are essential for designing. The research questions addressed in the first study are the following:

- What ranks are given by designers to the seven design tools for supporting the above knowledge processes?
- What features of the tools are responsible for influencing these rankings?

To study these processes, the structure of a piece of knowledge is described in terms of text and graphics. Textual knowledge is processed in terms of words and characters. Graphical knowledge is processed in terms of sketches, diagrams etc. Figure 12.1 shows the textual and graphical structure of an example drawing.

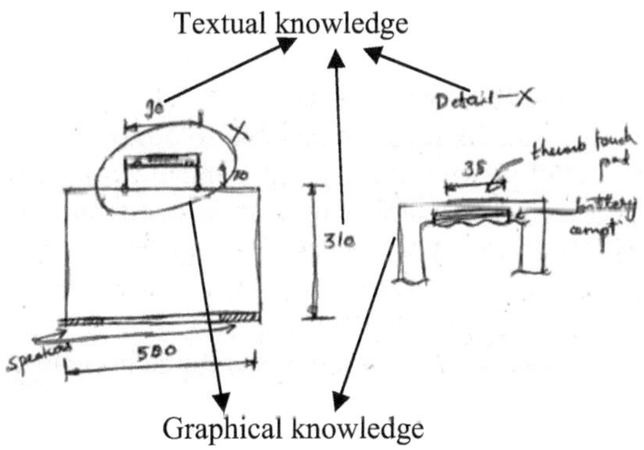

Figure 12.1. Textual and graphical structure of the knowledge

Table 12.1. Structure of initial experiments

Experiment	E_1	E_2	E_3	E_4	E_5	E_6	E_7	E_8	E_9	E_{10}	E_{11}	E_{12}	E_{13}	E_{14}
Designer(s)	D_1	D_2	D_1	D_2	D_1	D_2	D_1	D_2	D_1	D_2	D_1	D_2	D_1	D_2
Design Problem	P_1		P_2		P_3		P_4		P_5		P_6		P_7	
Equipment	Paper		Mobile e-Notes Taker™		Tablet without viewing facility		Tablet with viewing facility		Computer with Rhinoceros™ CAD		Mobilis™ PDA		P1i™ PDA	

For ranking the individual tools, two designers were asked to individually solve seven design problems, each using a different tool. Table 12.1 enlists the experiments conducted. The main purpose of the experiments was not to solve the design problems but to evaluate the tools in terms of the knowledge processes under focus. All the experiments were conducted in a single day. Before conducting the experiments, both designers were given adequate training to use the tools. After all the experiments were completed, the designers filled in a questionnaire (details in Ideas Lab - Questionnaire, 2010) to rank the tools on a scale of 1-7, with rank 1 representing the best tool.

12.3 Results and Discussion

The ranking given by the two designers for the individual tools are given in Figures 12.2-12.9. The main observations from these figures are the following:

- For generating textual and graphical knowledge, the designers rated Paper, Mobile E-Notes Taker™ and Tablet with viewing facility as the top three.
- For modifying/editing textual knowledge, Mobile E-Notes Taker™, Tablet with and without viewing facility, and Computer with Rhinoceros™ CAD package are ranked as the top three by both the designers.
- For modifying/editing graphical knowledge, Tablet with and without viewing facility and Computer with Rhinoceros™ CAD package are ranked as top three by each designer.
- For capturing textual knowledge, Mobile E-Notes Taker™, Paper, Tablet with and without viewing facility and Computer with Rhinoceros™ CAD package are the top choices. However, a major difference between the designers is observed in the ranking of Computer with Rhinoceros™ CAD.
- A major difference is observed in using Computer with Rhinoceros™ CAD for capturing graphical knowledge. This shows a difference in the designers' opinion about the use of this package in capturing knowledge.
- The pattern observed in reusing knowledge is similar to capturing knowledge. Once again, Mobile E-Notes Taker™, Paper, Tablet with and without viewing facility and Computer with Rhinoceros™ CAD package are ranked highly by both the designers.
- Giving equal weight to generation, modification/editing, capture and re-use of textual and graphical knowledge, the average ratings are: Mobile E-Notes Taker™ (2.6), Tablet with viewing facility (2.7), Computer with Rhinoceros™ CAD (2.8), Paper (3.4), Tablet without viewing facility (3.5), P1i™ PDA (5.9) and Mobilis™ personnel digital assistant (7).
- The reason why Mobile E-Notes Taker™ is ranked higher than the others is that it provides a blend of properties between paper and computer. Due to its properties similar to those of paper in generating textual and graphical knowledge it is ranked similar to paper in those respects, and due to its storage and editing properties similar to those of computer it has been ranked highly in modification, capture and re-use of knowledge.

- Note that Mobile E-Notes Taker™ is rated higher than Wacom™ tablet even though the latter has a larger working area and a greater resolution.
- Wacom™ tablet with viewing facility is rated higher than Tablet without viewing facility, despite the latter having higher pressure sensitivity levels.
- Compared to Mobilis™ personnel digital assistant, P1i™ PDA is rated higher despite the former having a greater working area and resolution. The important factor deciding this ranking in this case is the speed of response which depends upon the processor used in the system.
- Figure 12.10 shows the relationship between the ranking of the tools and their respective costs. Since the least and most expensive tools occupy the first and second rank, no common inference was possible to be drawn from this.

These observations stress the importance of features provided in the system for knowledge generation, modification, capture and re-use. Working area, resolution and pressure sensitivity levels did not seem to influence the designers' rankings much. This could be due to the nature of the design stages involved, requiring little precision and complexity to be handled. Based on the rankings, the three top tools (Mobile E-Notes Taker™, Tablet with viewing facility, and Computer with Rhinoceros™ CAD package) were selected for in-depth study, to understand the influence of the tools on knowledge capture and re-use during conceptual designing.

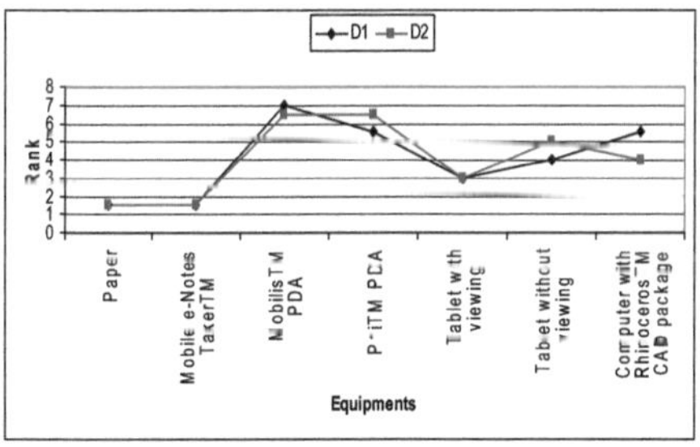

Figure 12.2. Generation of textual knowledge

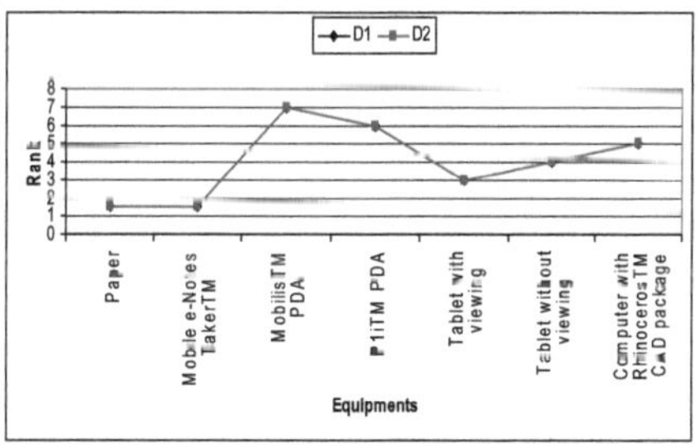

Figure 12.3. Generation of graphical knowledge

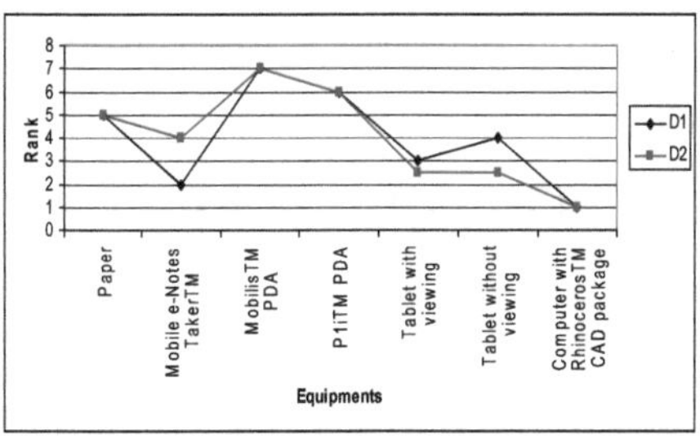

Figure 12.4. Editing textual knowledge

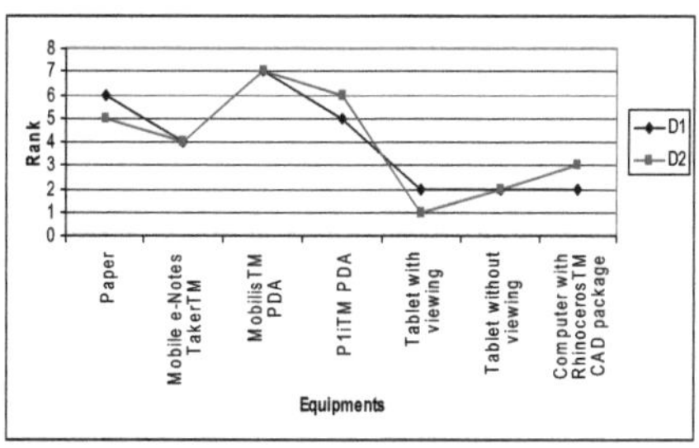

Figure 12.5. Editing graphical knowledge

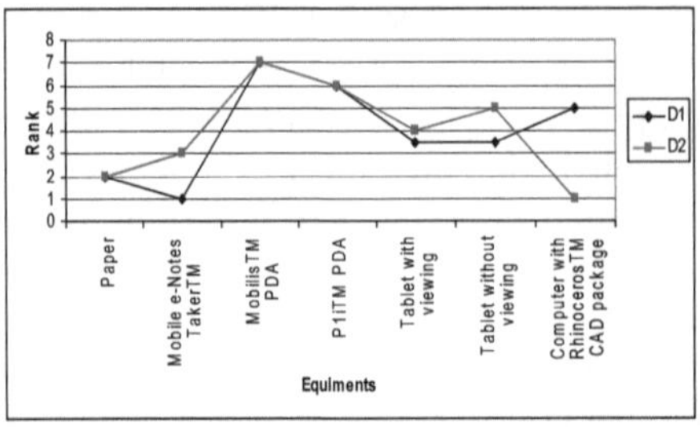

Figure 12.6. Capturing textual knowledge

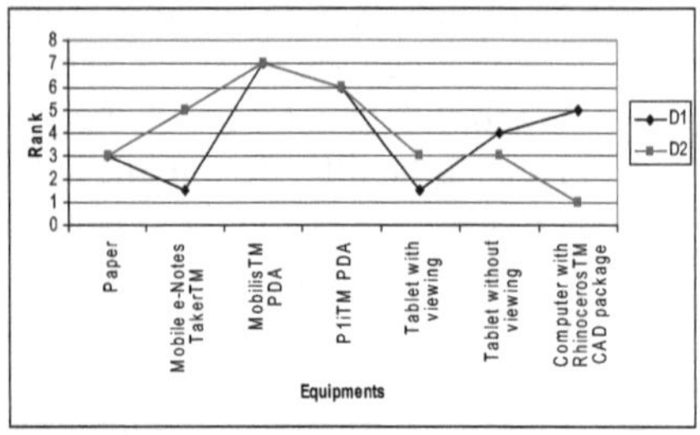

Figure 12.7. Capturing graphical knowledge

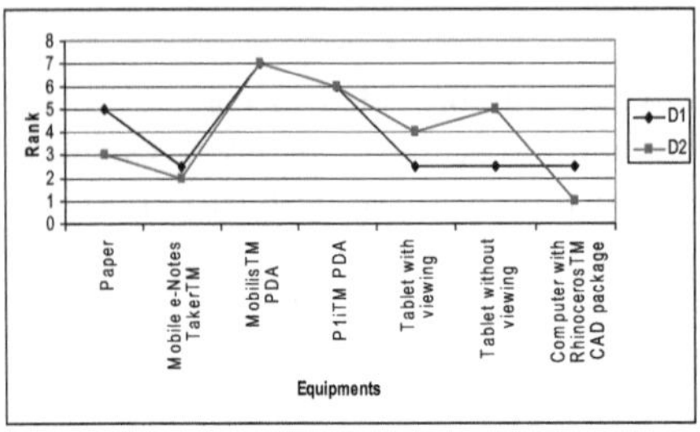

Figure 12.8. Reusing textual knowledge

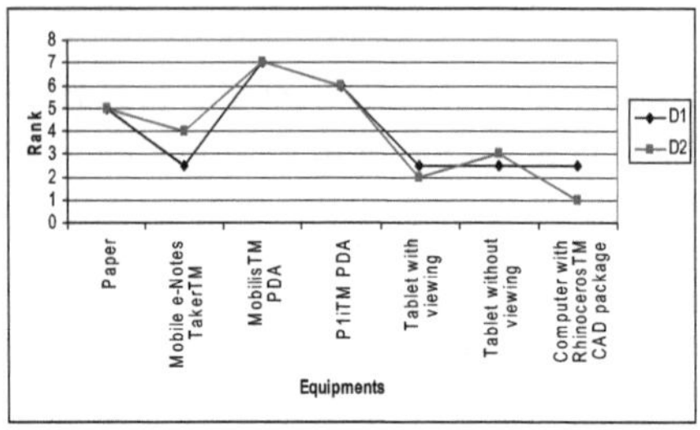

Figure 12.9. Reusing graphical knowledge

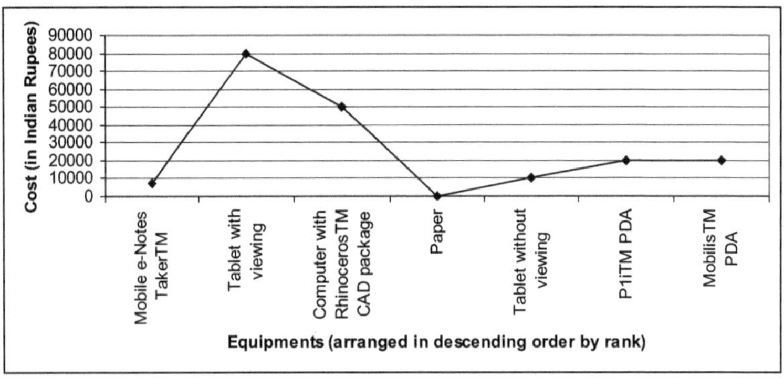

Figure 12.10. Equipment versus Equipment cost (1 British Pound ~ 75 Indian Rupees)

12.4 Second Study: Research Questions and Methodology

The objective of the second, in-depth study is to assess the specific impact of the three individual tools (Mobile e-Notes Taker™, Wacom™ Tablet with viewing facility and Computer with Rhinoceros™ CAD package on knowledge capture and re-use. The following research questions are asked:

- What knowledge do designers capture during designing?
- What knowledge do designers re-use during designing?

- How do designers perceive the characteristics of the captured knowledge during re-use?
- What are the impacts of the knowledge that is not captured during designing?

To answer these questions, in-house design experiments were conducted in a laboratory setting. To study the capture and re-use aspects, both original and redesign exercises were used. The experiments were structured taking into account the variations necessary to enable comparison of usage of the three tools, by multiple people, in both capture and re-use aspects. The structure of the experiments is presented in Table 12.2. Overall, 18 experiments were conducted with four Master of Design students and two design researchers, all with a prior degree in engineering (See details in Ideas Lab - Designers' details, 2010). Three different problems, (also detailed in Ideas Lab - Designers' details, 2010), were used across the 18 experiments. For the redesign experiments, documents captured during the original experiments were provided as an input. Each subject carried out task clarification and conceptual design to solve three different design problems, each with a different tool. Approximately one hour was given to a designer to solve a design problem. Designers were given adequate training to use the tools before conducting the experiments. During the experiments, each subject was asked to 'think aloud' such that the researcher could obtain a richer externalisation of their thoughts and activities during the experiments. All experiments were video-recorded and transcribed. Protocol analyses were carried out for each experiment with the help of a new taxonomy of knowledge proposed in our earlier work, the description and rationale of which has been given in (Gokula Vijaykumar and Chakrabarti, 2008). Definition of each term in the taxonomy is given with examples in (Ideas Lab - Taxonomy, 2010). Results obtained from these analyses are discussed in Section 12.5.

Table 12.2. Structure of experiments (D1-D6→Designers, E1 - E18 → Experiments)

	Problems (Original)			Problems (Redesign)		
Tools	**P1**	**P2**	**P3**	**P1'**	**P2'**	**P3'**
Mobile e-Notes Taker™	D6	D1	D5	D3	D4	D2
	E1	E2	E3	E10	E11	E12
Wacom™ Tablet with viewing facility	D2	D6	D3	D4	D5	D1
	E4	E5	E6	E13	E14	E15
Computer with Rhinoceros™ package	D1	D3	D4	D5	D2	D6
	E7	E8	E9	E16	E17	E18

12.5 Results

This section discusses the answers obtained for each research question from the analyses of the in-house design experiments.

12.5.1 What Knowledge Do Designers Capture During Design?

Protocol analyses were carried out to answer these research questions. The proposed taxonomy of knowledge described in (Gokula Vijaykumar and Chakrabarti, 2008) was used to code the protocols generated from the design experiments. Detailed percentages of knowledge captured in the nine experiments each for original and redesign experiments are tabulated in Ideas Lab - Detailed tables (2010). A summary of the results is given in Tables 12.3-12.4, for original and redesign experiments respectively. In the protocol, the number of lines generated and captured signifies the number of meaningful sentences uttered by the designer.

The observations from Tables 12.3-12.4 are as follows:

- The differences in the percentage of knowledge captured across the usage of the three tools indicate that tools may have an influence on the knowledge capture activity.
- Though none of the tools capture all the knowledge produced during designing, Mobile e-Notes Taker™ seems the best for capture, both in the original and redesign experiments. It is interesting to note that the price of Mobile e-Notes Taker™ is far below that of the other two tools compared.

Table 12.3. Summary table for original design experiments

Tools	Mobile e-Notes Taker™	Wacom™ tablet with viewing facility	Computer with Rhinoceros™ CAD package
	Average of respective three experiments		
Number of lines generated	222.3	137.7	191.7
Number of lines captured	75.7	46.7	41
% of capture	**40.5**	**35.8**	**23.4**

Table 12.4. Summary table for redesign design experiments

Tools	Mobile e-Notes Taker™	Wacom™ tablet with viewing facility	Computer with Rhinoceros™ CAD package
	Average of respective three experiments		
Number of lines generated	86	188.3	163.7
Number of lines captured	49	59.3	59.3
% of capture	**54.7**	**27.9**	**35.1**

12.5.2 What Knowledge Do Designers Re-use During Design?

The total extent of observable knowledge in these experiments is the summation of the extent of knowledge related to 'verbalisation of designer's thoughts' and 'documents captured on computer'. The focus to answer the re-use questions is based on the documents captured on computer. Knowledge captured in the documents was used thoroughly by the designers in all the nine experiments, irrespective of the tools used. Designers spent adequate time to understand the requirements and solutions generated by the previous designers. This shows that designers are willing to get as much as information and knowledge as possible from the stored documents. Only few cases were found where a designer did not explore a concept properly. Reasons for this are given in Section 12.5.3.

12.5.3 How Do Designers Perceive the Characteristics of Captured Knowledge During Re-use?

The aim of this question is to understand whether the quality of captured knowledge provided to the designers is adequate. The criteria used to assess quality are: accuracy (nearness to the truth or true value), clarity (ease of understanding) and completeness (presence of everything required). It is interesting to note that in most cases designers did not comment negatively about the captured knowledge. This indicates that designers accept the given captured knowledge as it is. The problems that occurred due to improper capture are discussed in the next section.

12.5.4 What are the Impacts of the Knowledge that is not Captured During Designing?

The video recordings of the experiments were transcribed, and each transcription was divided into events. Each event from the design experiments was coded using the following categories: 'captured knowledge' and 'not captured knowledge'. The impacts of these 'not captured knowledge' events were analysed by matching events held in redesign experiments. It has been shown in Tables 12.3-12.4 that by using the various tools, designers roughly captured about 23% to 40% of the knowledge produced during original designs. The impact of the remaining knowledge that is not captured during the original design was analysed. The analyses were performed by comparing the protocols generated from the respective original and redesign experiments. The problems during redesigning that were identified are as follows:

- One designer stressed the need of a product description after visualising the structural knowledge from Rhino^{TM} software.
- Four designers regenerated requirements that were not captured in the original experiments.

- Two designers ignored some concepts from the original design because the behaviour of these concepts was not known (from the original design experiments).
- Two designers attempted to understand the original concept by querying to understand the thoughts of the designer involved in the original design experiment. This means that they would have benefited from the missing original knowledge.
- Three designers ignored those features in the original designs for which adequate knowledge was not captured and provided.
- Two designers misinterpreted the original concepts due to partially captured knowledge.
- Three designers did not explore those original concepts which they could not understand.
- Four designers made many assumptions about the concepts by stating that the original designer might have thought about these but did not capture.
- Two designers expressed a concern by stating whether their handwriting could be understood by others.
- One designer reworked on the missing and partially captured concepts.
- One designer missed requirements that were not properly written.
- Two designers ignored concepts that were not clear to them.
- Three designers seemed to be satisfied with a partially understood concept. Due to this partial understanding they made wrong decisions in redesigns.
- Two designers wrongly assumed a feature due to missing knowledge.

12.6 Discussions and Conclusion

From the literature survey, it was concluded that even though many tools are developed to support knowledge capture and re-use, their effectiveness have not been studied in-detail, especially from the point of view of knowledge processes. The aim of this paper is to analyse the effectiveness of various tools for knowledge capture and re-use, in order identify or develop tools to better support designers to satisfy their knowledge needs. Use of these tools during design should increase the effectiveness of interactions carried out by designers for capture and re-use.

Evaluations of seven individual tools for generation, modify/edit, capture and re-use processes reveal that:

- Designers in some cases have varying opinions about which tools are better for capturing knowledge.
- Mobile E-Notes Taker™, Tablet with viewing facility and Computer with Rhinoceros™ CAD package are rated as the top three tools for the processes considered.
- It could be argued that Mobile E-Notes Taker™ is rated highly because it combines the metaphors of both paper and computer.
- It is noted that the size of the working area, resolution and pressure sensitivity does not play an important a role in these processes during task

clarification and conceptual design stages, unlike what was initially expected by the researcher.

- The speed of response plays a vital role in deciding the rankings between the tools considered.

Detailed evaluation of the three top rated tools reveals that:

- Tools definitely influenced designers' behaviour in capturing knowledge.
- None of the tools were adequate in capture of all the generated knowledge.
- The influence of Mobile E-Notes Taker™ on capturing knowledge is higher than that using the other two, more expensive, competing tools.
- The analyses do not provide clarity on the different types of knowledge that are captured by the designers.
- During redesigning, the designers spent adequate time to understand the requirements and solutions generated by the previous designers. This shows that designers are willing to get as much as information and knowledge from the stored documents, if original designer is not available.
- It is likely that designers will accept the given captured knowledge as is for re-use, if the designer who produced that knowledge is not available.
- Various problems were found to have been encountered by the designers, as observed in the protocols, due to the knowledge that was not at all, partially, or improperly captured.

These results show that the tools have a varying influence on the capturing effectiveness; but still none of the tools used is adequate. Designers do not intuitively capture all the knowledge required for re-use, and designers' behaviour for capturing knowledge is highly unpredictable. This suggests that some other mechanisms should be added to these tools so that their effectiveness of knowledge capture could be increased. One mechanism is to incorporate and stress the importance of interactions during the design process, and the use of a taxonomy of knowledge during designing. As part of our research, we have developed the KRIT model (acronym for Knowledge of solutions - Requirements - Interactions and Tasks) of interactions between knowledge elements and their influences (detailed in Gokula Vijaykumar and Chakrabarti, 2006). Incorporating this model and the proposed taxonomy of knowledge through these equipment during the design process is a current focus of our work. This study has been conducted with relatively novice designers (note that they all are graduate engineers with some experience of designing), with the hope that this can be validated and further extended with more experienced designers. It should be noted that the three tools evaluated in detail were evaluated only when designers worked individually, that too for task clarification and conceptual design stages only. Presently this is being further extended by analysing other design experiments carried out by design teams, and in both collaborative (where designers work together in the same place and time) and distributed (where designers work together in the different place and same time) set-ups. Our current study investigates influence of features of design tools on knowledge capture and re-use. Also knowledge articulation between textual and graphical parts will be studied in our future work.

12.7 References

Cugini J, Damianos L, Hirschman L, Kozierok R, Kurtz J, Laskowski S *et al.* (1999) Methodology for evaluation of collaborative systems. The Evaluation Working Group of The DARPA Intelligent Collaboration and Vizualisation Program, The MITRE Group, Bedford, MA, US

Mugellesi-Dow RM, Pallaschke S, Merri M, Montagnon E, Schabe M, Belingheri M *et al.* (2008) Overview of the knowledge management system in ESA/ESOC. Acta Astronautica 63(1-4): 448-457

Duffy AHB, Duffy SM (1996) Learning for design reuse. International Journal for Artificial Intelligence for Engineering Design, Analysis and Manufacture (AI EDAM) 10(2): 139-142.

Encore (2008) Available at: http://www.ncoretech.com/products/ia/mobilis/index.html (Accessed on 13 February 2009)

Gokula Vijaykumar AV, Chakrabarti A (2006) Understanding of knowledge generation during design process in industry. In: Proceedings of the International Conference on Product Life Management (PLM'06), Bangalore, India

Gokula Vijaykumar AV, Chakrabarti A (2008) Understanding the knowledge needs of designers during design process in industry. Journal of Computing and Information Science in Engineering 8(1): 011004-1--011004-9

Hi-Tech Solutions (2008) Available at: http://www.hitech-in.com/mobile_e-note_taker.htm (Accessed on 13 February 2009)

iBall (2008) Available at: http://iball.co.in/Product.aspx?c=16 (Accessed on 13 February 2009)

Ideas Lab - Designers' details (2010) Available at: http://cpdm.iisc.ernet.in/ideaslab/ Problems used in assessing design tools.mht (Accessed on 2 January 2010)

Ideas Lab - Detailed tables (2010) Available at: http://cpdm.iisc.ernet.in/ideaslab/Detailed tables for percentages of knowledge captured.mht (Accessed on 2 January 2010)

Ideas Lab - Equipment (2010) Available at: http://cpdm.iisc.ernet.in/ideaslab/Specification of equipments.mht (Accessed on 15 January 2010)

Ideas Lab - Questionnaire (2010) Available at: http://cpdm.iisc.ernet.in/ ideaslab/Questionnaire for assessing design tools.mht (Accessed on 15 January 2010)

Ideas Lab - Taxonomy (2010) Available at: http://cpdm.iisc.ernet.in/ideaslab/Taxonomy of Knowledge.mht (Accessed on 2 January 2010)

Pinelle D, Gutwin C (2000) A review of groupware evaluations. In: Proceedings of the 9th IEEE International Workshops on Enabling Technologies: Infrastructure for Collaborative Enterprises (WETICE'00), Washington, DC, US

Rhinoceros (2008) Available at: http://www.rhino3d.com/ (Accessed on 13 February 2009)

Sony Ericsson (2008) Available at: http://www.sonyericsson.com/cws/corporate/products/ phoneportfolio/specification/p1i (Accessed on 13 February 2009)

Taylor RS (1996) Value-added processes in information systems. Ablex Publishing, Norwood, NJ, US

Vizcaíno A, Piattini M, Martinez M, Aranda GN (2005) Evaluating collaborative applications from a knowledge management approach. In: Proceedings of the 14th IEEE International Workshops on Enabling Technologies: Infrastructure for Collaborative Enterprise (WETICE'05), Linkoping, Sweden, pp 221-225

Wacom (2008) Available at: http://www.wacom.com.au/previous/cintiq/DTU-710/DTU-710.html (Accessed on 13 February 2009)

Part IV

Engineering Process Management in Practice

Chapter 13

Modelling the Quality of Engineering Designs

W.P. Kerley and P.J. Clarkson

13.1 Introduction

13.1.1 Background

In the turbulent environment that characterises business today, companies are under constant competitive pressure to develop innovative, high quality products in increasingly shorter timescales. Companies that want to remain competitive are continuously trying to better manage and improve their product development (PD) processes. A key element of PD process management and improvement is to understand and measure PD processes and their outcomes (O'Donnell and Duffy, 2005). These measures traditionally involve an evaluation of both the efficiency of the process - process time and spend - and the effectiveness of the process in terms of the quality of the output *e.g.* product performance.

There is a growing body of literature concerning the use of activity-based process modelling as a tool for supporting the management and improvement of PD processes. Activity-based models decompose processes into a network of discrete, interlinked tasks and are commonly represented as either flowcharts or dependency structure matrices (DSMs) (see Browning and Ramasesh, 2007 for a recent review). This literature overwhelmingly uses reduction in process time as the basis for evaluating process improvement. At the same time, the comparatively small number of studies which model measures of product performance in PD has been highlighted by a number of authors (*e.g.* Browning and Ramasesh, 2007; Eriksson and Brannemo, 2009). A bias towards modelling measures of efficiency over those of effectiveness seems problematic given the trade-offs between time, spend and quality that occur in real-life practice. Ideally, if process modelling is going to be used to support PD process management and improvement, the models used should incorporate measures of process time and spend, product quality and the relationship between them (O'Donnell and Duffy, 2005).

13.1.2 Motivation and Purpose

This paper is motivated by ongoing research with Rolls-Royce Civil Aerospace which uses activity-based process modelling to propose improvements to the company's early stage design processes. The project the authors are currently working on is developing proposals for how to further automate conceptual engine design activities and involves greater use of computational modelling tools and increased integration between these tools. The purpose of this process modelling is to quantify the value that will be derived from this automation. Ideally this will allow comprehensive comparisons to be made between the "as-is" model of the current process and the "to-be" model of the proposed process along the dimensions of time, spend and quality of the preliminary design, allowing for the trade-offs between these different performance measures. However, there are a number of gaps in our knowledge to be bridged before this is possible.

While performance measurement has been well researched for manufacturing (see Neely, 2005 for a review), a comprehensive, canonical set of process measures across the whole design and development process does not exist. Although measurement of PD process time and spend, as well as the quality of manufactured products at the end of the PD process, are reasonably well understood, there seem to be at least two areas where more research is required: (1) how to measure the quality of intermediate, largely information-based deliverables in the development process, such as engineering designs; (2) how to model the influence of the PD process on the quality of its outputs. The purpose of this paper is to consider the current state of knowledge related to measuring and modelling engineering design quality using activity-based models and to suggest possible research directions going forward. Relevant research questions include, but may not be limited to:

- What characterises the quality of an engineering design?
- How can the quality of an engineering design be measured and evaluated?
- How does design quality develop over time during the design process?
- Which organisational and process factors determine the quality of an engineering design at any stage in the design process?
- How can measures of design quality be incorporated into activity-based process models?

All of these questions have, to some degree, been addressed in the literature; although, as far as the authors are aware, the various strands of literature have not been pulled together for the purpose of modelling the quality of engineering designs. This paper represents preliminary thoughts on this subject. It does not attempt to systematically and comprehensively review the entirety of this literature: such an undertaking is beyond the scope of - and space allowed by - a conference paper such as this. Instead it builds upon the authors' research in the aerospace sector, particularly related to the design of civil jet engines. In this sense the paper is biased to aerospace issues and neglects a number of factors that are relevant in other engineering design domains. However, it does give an overview of the modelling issues involved, highlights some of the key literature that addresses these issues and provides suggestions for how this research could be taken forward.

13.1.3 Paper Structure

The organisation of the remainder of this paper is as follows. Section 13.2 considers engineering design and looks at what is known about design quality and how it is ensured in practice. Section 13.3 considers what the implications of this practice are for creating an activity-based model of the PD process. This analysis raises a number of requirements that such a model should address if process time and spend, product quality and the relationship between them are to be modelled. Section 13.4 then discusses the existing literature related to the points previously raised and highlights opportunities for further research. Section 13.5 concludes the paper by summarising the key points made and outlining how the authors expect to take this research forward.

13.2 Engineering Design Practice

13.2.1 Design Quality

An engineering design (the artefact) is a specification for a product or part of a product. Engineering designs are built up in stages of increasing detail, for example progressing from conceptual design through embodiment design to detailed design (Pahl and Beitz, 1996), until there is sufficient detail for the product to be manufactured as intended. If the manufactured product is to be successful then it must deliver customer value, which depends on its affordability (cost), lead time and performance. Engineering designs produced during product development should contribute information to realise this product value (Browning *et al.*, 2002).

The value of any artefact is a function of benefits gained from it compared to the cost of obtaining those benefits. For an engineering design, the costs involved are the time and resources used to produce that design. The benefits are a function of the intrinsic quality of the design and its relevance to the remainder of the PD process. Design quality is comprised of a number of factors (Browning *et al.*, 2002; Grebici *et al.*, 2008a). These include:

- *The expected product performance.* What will the performance of the finished product be? This is usually considered in terms of conformance to the requirements that have been set for the design.
- *The information content.* What is the scope of the design? How detailed is the design specification? How much more work is needed to complete it?
- *The uncertainties and risks associated with the design.* What is the likelihood that this design will develop into the desired final product within the expected development time and spend?

Considering the last point in more detail, there are many sources of uncertainty. According to Eversheim *et al.* (1997) these uncertainties fall into two categories: uncertainty of content and uncertainty of context. Uncertainty of content is made up of imprecision, incompleteness and vagueness; to which could be added errors (Grebici *et al.*, 2008a). Uncertainty of context is made up of reliability, variability and validity of the information. The sources of these uncertainties vary but, according to Grebici *et al.* (2008b), can come from: 1) the interdependencies arising from design decomposition and parameterisation; 2) uncertainties arising from multiple descriptions of parameters; and 3) uncertainties arising from the process of developing the design. As a design develops over time its maturity increases: more information is produced about the design and the knowledge gained from increased information is expected to reduce the risks associated with these uncertainties (Browning *et al.*, 2002).

The impact of a specific design's quality on overall product development is dependent on its relevance and importance to the remaining PD activities. Impact therefore depends upon (Grebici *et al.*, 2008a): 1) the information quality in the design, which includes its usability, currency, accuracy and whether it is trusted (Zhao *et al.*, 2008); 2) the knowledge and experience of the users of the design; and 3) the sensitivity of later work to the uncertainties in the design, in particular the impact of having to rework later work if the design changes as a result of the uncertainties associated with it (as described above).

To recap, the value of a design is a measure of its benefits compared to its costs. The benefits of a design are a function of the quality of the information that it contains, the relevance of this information to its recipients, uncertainties associated with the information and the risks to the project that these uncertainties entail based on their likelihood of causing an unexpected event later on and impact of such an event occurring. Its costs are the project time and spend needed to create it, including all iterations and rework. Understanding and managing the value of intermediate designs is particularly pertinent to the design of complex products, where designs are typically produced and used by different teams both within the same company and across supply chains, and is consistent with the total quality management (TQM) principle of focusing on the needs of the "customer", wherever they are in the organisation.

13.2.2 Designing

To create an engineering design, there is a recognisable design process involved in product development, which can be broken down into a set of interrelated design activities (Browning and Ramasesh, 2007). For complex systems, such as a civil jet engine, the design is built up over time in a number of identifiable stages by a large number of engineers organised into various teams working concurrently on specific subsystems and/or components. It may take many years from the first conceptual designs being proposed to a fully realised product being manufactured.

Conceptually, at any level of detail, design involves analysis of requirements, synthesis of the design and evaluation of that design. For an engineering design the requirements are specified as a series of performance attributes. For example the headline technical performance requirements for a civil jet engine are: thrust, fuel consumption, weight, noise and emissions; in addition the engine must be delivered on time at an appropriate cost (Kirk, 2003). Design *analysis* involves understanding these requirements. Design *synthesis* takes these requirements and creates a design which can be described in terms of design parameters *e.g.* physical dimensions, materials *etc.* Design *evaluation* takes these parameters and uses them to determine the estimated performance attributes of the design for comparison with the original requirements. The generation of the estimated performance attributes often makes use of computational modelling such as computational fluid dynamics (CFD) and finite element analysis (FEA). Repeated iteration of the three steps - analysis, synthesis, evaluation - will typically be required to find a design that meets the requirements, is consistent with external constraints and has the best balance between performance characteristics and potential risks carried forward into the remainder of the PD process.

13.2.3 Design Management

There may be difficulty accurately estimating and scheduling PD projects, particularly given the highly iterative nature of PD, but there are concrete measures for how long a project has taken (time) and how much resource has been committed to it (spend) so far. Unlike manufacturing, no similarly "hard" measures exist for design quality and risks during PD (see Brookes and Backhouse, 1998 who provide reasons for this). Consequently, the evaluation of the risks associated with a particular engineering design, particularly at the very early stages of a PD project, is highly subjective due to lack of metrics and the inherent uncertainties involved. Projects rely heavily on the experience of the design team and the confidence they have in the design proposal, the evaluation data that they have available (*e.g.* from computer modelling or prototype testing), the specific designers involved and the ability of the project to further develop this design as intended. There are a number of methods and tools that are employed to do the design and to support risk evaluation. These include checklists, design reviews, failure modes and effects analysis (FMEA), fault tree analysis (FTA), quality function deployment (QFD), TRIZ and process modelling.

A (small) number of authors (*e.g.* Paquin *et al.*, 2000; Browning *et al.*, 2002) have proposed methods for systemising and increasing the quantification of risk analyses. Nevertheless the lack of objective, quantified measures of design quality and risk remains an impediment to design management and in particular to the modelling of the design process for process improvement purposes.

13.3 Modelling

13.3.1 Background and Case Example

There are various ways that process models can be produced. The main approach used in engineering design is activity-based modelling, which conceptualises a project as a process, decomposed into a network of activities (Browning and Ramasesh, 2007). Activity-based models can be documented using a number of different methods and notations, and analysed in a number of ways, including computer-based simulation. The authors have been using the Applied Signposting

Method (ASM) (Wynn *et al.*, 2006) at Rolls-Royce. It is, however, acknowledged that there are other capable methods and tools that are available within the engineering design community and that could be used for this purpose.

The current problem being considered at Rolls-Royce involves the automation of the early stages of product design. Rolls-Royce expects there to be significant improvements in process time and spend as well as improvements in design quality, given that design work will be done earlier during product development thereby reducing later development risks. Currently (January 2010), the modelling is at an early stage and the research questions posed in this paper are still to be resolved. Nevertheless it is clear that the final model will be large (>500 tasks) and highly complex, given the process involves many discrete, concurrent, interrelated activities and large amounts of iteration. A fragment (about 5%) of the process model is shown in Figure 13.1 to give an indication of this complexity and the challenges involved in using it to predict the impact of the proposed changes.

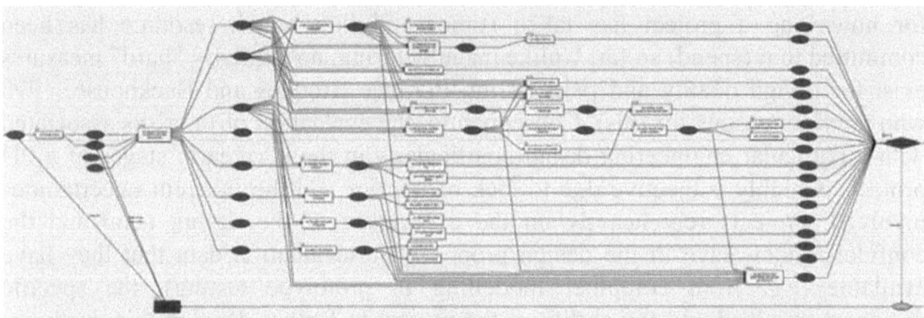

Figure 13.1. A model fragment of the Rolls-Royce preliminary design process for a civil jet engine (with the task and parameter names disguised) illustrating the complexity involved

13.3.2 Modelling Requirements

There are a number of requirements for building process models that can be used to support process improvement interventions such as at Rolls-Royce. At the highest level, first, in order to fully compare the performance of "as-is" and "to-be" process, you have to: 1) analyse design (artefact) performance; 2) analyse design activity performance (time and spend); and, 3) *relate the two* (O'Donnell and Duffy, 2005). Second, in order to do a computer-based simulation these performance variables have to be quantified. There are many cases of subjective, ordinal values (*e.g.* high, medium, low) being used for analysis. The ideal would be objective, absolute values as this would greatly increase models' analytical power. Finally, decisions have to be made about the level of abstraction in the model. Any model is an abstraction of reality and it is sensible to limit the number of variables modelled in order to manage model size (Kellner *et al.*, 1999); limiting the number of variables: reduces the amount of data that needs to be gathered to populate the model, abstracts away the detail for comparison purposes; and generalises the model for use across multiple PD projects.

At the detailed level there are a number of issues to be addressed that can be explained with reference to a conceptual framework for design practice (see Figure 13.2 below). The main elements of this framework, discussed in Section 13.2 above, are: the design (artefact), designing and design management. Challenges identified that need to be overcome in order to produce a simulateable model are represented as numbers on the figure and are explained below.

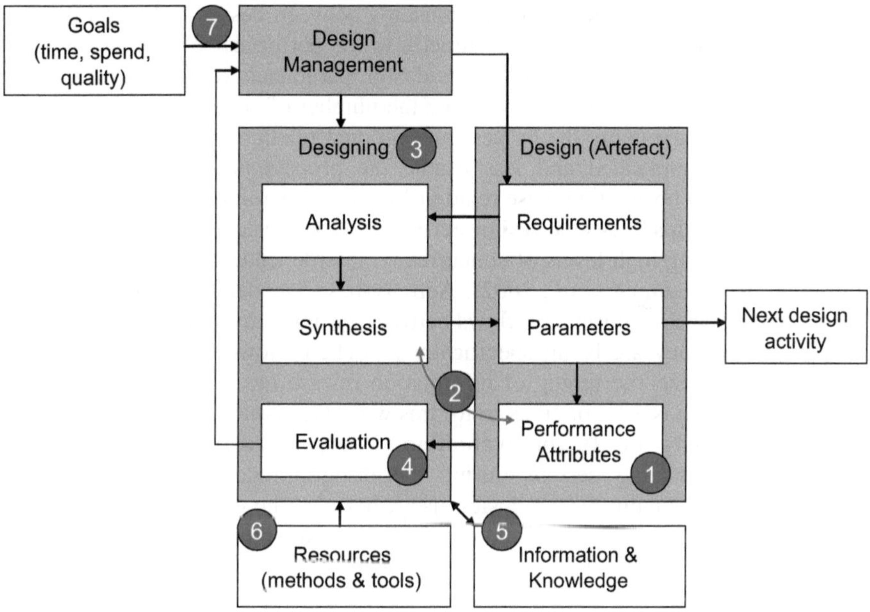

Figure 13.2. Conceptual model of the design process

1. The number of performance attributes. A complex product, such as a jet engine, is composed of thousands of different components each with its own performance attributes. Some attributes, such as weight and unit cost, are relevant to all components and others, such as aerodynamic properties, are not. The various performance attributes are part of a hierarchy, cascading down from those of the whole product to those of individual components. In practice, performance attributes at different levels in this hierarchy are considered at any time depending on the purpose of the design evaluation.

At Rolls-Royce, there are five highest-level performance attributes: weight, fuel consumption, turbine entry temperature, cost and noise. These represent the minimum set to be modelled if trade-offs between attributes are to be considered. At the same time, modelling at the next level down leads to a significant increase in the number of attributes involved, as well introducing extra complexity required to resolve the relationships between performance attributes across and between different levels in the performance attribute hierarchy. (see, for example, Ghisu *et al.*, 2008 concerning gas turbine compressor optimisation).

2. The relationship between design activities and performance attributes. The relationships between design activities, design parameters and performance attributes are complex. In most PD process models the relationship between the process and the product is not explicitly shown, although implicit in the process structure. That said, certain modelling methods, such as OPM (Dori, 2002) and SysML, do explicitly include the relationships between design tasks and product features; while others link the tasks directly to the performance attributes without specifically modelling the product or its design parameters (*e.g.* Browning *et al.*, 2002). Like the previous issue concerning the number of performance attributes to model, the more explicitly the process and product relationships are modelled the bigger, more complicated and less usable the process model will become. A suitable means of abstracting these relationships may, therefore, be required.

3. Relationships between design activities. Most design processes are highly complex, involving high levels of concurrency, interdependency between activities and iteration (Browning *et al.*, 2002). Activity-based models tend to simplify the representation of the information flows between design activities unless the effects of task concurrency are being specifically modelled. However, as Grebici *et al.* (2008a) make clear, the timing of information release has a significant effect on design risk, which is a major concern for this work (see issue 7 below).

4. Subjectivity in design evaluation. There is considerable subjectivity involved in evaluating designs during the design process. These evaluations consider both the quality of the design itself, but also the expected impact of this design on later design activities, and rely heavily on the previous experience of the design team. How can, and should, the concept of "confidence" in a particular design be modelled? In particular, what criteria should be modelled for design evaluations to determine whether the model will: 1) progress to the next activity; or 2) iterate the current design activity to increase confidence before proceeding?

5. Understanding and reflecting design context. The value of information used throughout the design process varies based on the context of its use and the knowledge of the individuals using it (Zhao *et al.*, 2008). The novelty of the product being designed and the skill and experience of the designers are, therefore, both factors affecting the efficiency of the process and the quality of the design produced. This context is often not considered when process modelling, but it would seem necessary, if a model is going to be generalisable to both routine and novel product design processes.

6. Effect of different resources. In addition to the people allocated to a design activity, the methods and tools they use, and how well they use them, will affect the efficiency and effectiveness of the design process (O'Donnell and Duffy, 2005).

7. Metrics and trade-offs. Most design process models use process time to measure process performance. Activity times may be modelled deterministically or stochastically depending on the model application. Process spend can be derived from process time based on resource allocations to particular activities. There is less clarity on which aspects of design quality to model and how to measure these. The authors' view is that the key quality metric should be the risk carried forward to subsequent design activities, which could be quantified in terms of either time or money. Moreover, we would like to be able to model the impact of varying time (*e.g.* changing the time spent on specific activities or changing the process structure) and resource allocations on this design risk. In general, all other factors being unchanged, reducing the time spent on a task would be expected to reduce the quality of the design produced and the risks carried forward to later stages of a project, but what about process improvement approaches such as task concurrency or process automation? Trade-off surfaces could assist in the evaluation of these types of process change and in the identification of "sweet spots" between time, spend and quality for particular design process configurations.

13.4 Discussion

Having reviewed engineering design practice and the issues of using activity-based modelling to model this practice, we return to the six questions posed in Section 13.1.2 and look at the state of research within the Engineering Design community related to these questions, the implications for the work with Rolls-Royce, and possibilities for further research.

What characterises the quality of an engineering design? This question was addressed in Section 13.2.1. There is a significant body of literature that either deals directly with this topic (*e.g.* Aas *et al.*, 1992; Paquin *et al.*, 2000) or looks at specific aspects of it, in particular uncertainty and risk (*e.g.* Browning *et al.*, 2002; Grebici *et al.*, 2008a). Other work being done at Bath and Loughborough, which considers the value of information (*e.g.* Zhao *et al.*, 2008), is also relevant as a design is an information artefact. Overall there seems to be a good, and ever increasing, understanding of what factors affect design quality.

How can the quality of an engineering design be measured and evaluated? The lack of hard, quantitative measures for design quality was considered in Section 13.2.3. Work on measuring and auditing design risk includes Browning *et al.* (2002). This work is useful because it aims to quantify these risks. On the other hand, the measures are estimates of final product performance based on the design to date; for process improvement purposes, such as the work with Rolls-Royce, leading and, ideally, objective measures of intermediate design quality are required. The issue of measurement is fundamentally that design is an information artefact and information is difficult to measure. Zhao *et al.* (2008) have developed a method for calculating information value using a Bayesian network approach. More work will be needed before this can be incorporated into the Rolls-Royce model given: 1) the desire to limit the number of variables used to model design quality; and 2) the need to incorporate risk and uncertainty. Grebici *et al.* (2008a) start to address the second of these two points by modelling information passing between pairs of design tasks and it may be possible to extend their method to multi-task process models (Grebici *et al.*, 2008b), such as at Rolls-Royce.

The other strand of research is to look at how design quality is evaluated in practice. This is largely the realm of the performance management literature, which is strong for manufacturing (Neely, 2005) and less developed in engineering design; however work is ongoing at various universities (*e.g.* O'Donnell and Duffy,

2005; Johansson *et al.*, 2009; Johnsson *et al.*, 2009) in this field. For our work, development of new measurement systems is less interesting than understanding how and why particular decisions are made as to whether to let a project proceed or to iterate design activities again. Further descriptive case studies are required of actual practice and how confidence in a particular design is gained.

How does design quality develop over time during the design process? A theme in the design literature is that of maturity. The principle is that over time more information about the design is created, increasing certainty about the final product specification and thereby reducing the number and severity of project risks (Browning *et al.*, 2002). It should be noted, however, that the quality of the design does not increase monotonically with time and there can be periods where little apparent progress is made followed by surges where higher product performance is achieved (Jarrett and Clarkson, 2002). This causes problems for modelling given that it complicates any trade-off functions between time, spend and quality.

Which organisational and process factors determine the quality of an engineering design at any stage in the design process? Besides time, including iterations and rework, and spend, the quality of the design will also be determined by the resources - people, methods and tools - allocated to it. There is a strong literature on engineering design methods and tools and their effects on project performance (*e.g.* Schabacker, 2002; O'Donnell and Duffy, 2005), although this lacks direct causal links to design quality, which is required for the predictive modelling we want to do. The impact of people on the process has not arisen so far as a major issue in our work at Rolls-Royce, nor in the literature reviewed for this paper. Nevertheless this is a major theme in other fields such as software engineering, and therefore research is known to be available if it is found to be required at a later stage.

How can measures of design quality be incorporated into activity-based process models? This question was elaborated in Section 13.3.2. There is a very broad body of literature related to PD process modelling in general (Browning and Ramasesh, 2007). However this literature is fragmented with respect to the problem of modelling design quality using activity-based models.

The majority of the activity-based modelling literature uses time as the process performance measure. Within this literature, the work that models overall process scheduling allowing for concurrency and design iteration (see Browning and Ramasesh, 2007) is particularly relevant to the Rolls-Royce work (and the problem of modelling design quality in general), given that these factors play such an important role in actual practice. There is a more limited literature that considers aspects of design quality (*e.g.* Clarkson and Hamilton, 2000; Browning *et al.*, 2002; Grebici *et al.*, 2008a; Lévárdy and Browning, 2009); which can be built upon.

The main problem with modelling design quality at the same level as process time and spend is that it relaxes more of the constraints and, therefore, increases the number of variables for consideration. This significantly increases the modelling complexity involved. A suitable approach needs to be determined that balances fidelity with usability and leads to new methods, and possibly new tools.

Coming back to the points raised in Section 13.3, we also still need to find an objective scheme for quantifying the value/quality of a design before, during and after the design process. Given that this means quantifying information, there are fundamental epistemological questions that have to be addressed. In particular how does knowledge (rather than information) about the design build up over time and how does this knowledge affect decision making during the design process, for example during design evaluations and at process stage gates.

13.5 Conclusion and Next Steps

Modelling design quality is a real challenge for companies, like Rolls-Royce, who develop complex products and want to use process modelling to improve their design processes. This paper has laid out the problems involved and reviewed the basis of possible solutions. In the absence of extant research in the areas listed above, the authors intend to take a pragmatic approach to modelling design quality for Rolls-Royce, making best use of the research that is currently available.

The first thing to do is to model quality and risk at the decision points in the process model only (rather than trying to calculate the contribution of all the tasks to design quality) as this is needed to determine iteration probabilities (and is a direct extension of time and spend modelling). To do this will involve understanding better how evaluation decisions are made and determining the most appropriate means of modelling these. This will probably be stochastically, based on a simple time and risk trade-off function.

The second stage will probably be to look at a single performance attribute and then trace its development through the whole process. This will require some sort of calculation at each stage. This complicates modelling as it requires a much better model of the linkage between process and product, and may require the use of other methods such as OPM or SysML. On the other hand, the model may be simplified by only considering the deltas between as-is and to-be processes.

In the meantime we hope that the rest of the engineering design community continues to develop their own research that will feed into this work. Currently, the work being done on valuing information and quantifying risk seems most complementary to our work concerning the mechanics of process modelling itself.

13.6 References

Aas EJ, Klingsheim K, Steen T (1992) Quantifying design quality: A model and design experiments. In: Proceedings of EURO-ASIC'92, Paris, France, pp 172-177

Brookes NJ, Backhouse CJ (1998) Measuring the performance of product introduction. Journal of Engineering Manufacture 212(1): 1-11

Browning TR, Deyst JJ, Eppinger SD, Whitney DE (2002) Adding value in product development by creating information and reducing risk. IEEE Transactions on Engineering Management 49(4): 443-458

Browning TR, Ramaseh RV (2007) A survey of activity network-based process models for managing product development projects. Production and Operations Management 16(2): 217-240

Clarkson PJ, Hamilton JR (2000) 'Signposting', a parameter-driven task-based model of the design process. In: Research in Engineering Design 12(1): 18-38

Dori D (2002) Object-process methodology: A holistic systems paradigm. Springer-Verlag, New York, NY, US

Eriksson J, Brannemo A (2009) Decision-focused product development process improvements. In: Proceedings of the 17th International Conference on Engineering Design (ICED'09), Stanford, CA, US

Eversheim W, Roggatz A, Zimmermann H-J, Derichs T (1997) Information management for concurrent engineering. European Journal of Operational Research 100(2): 253-265

Ghisu T, Parks GT, Jarrett JP, Clarkson PJ (2008) Integrated Design Optimisation of Gas Turbine Compression Systems. In: Proceedings of the 4th AIAA Multidisciplinary Design Optimization Specialist Conference, Schaumburg, IL, US

Grebici K, Goh YM, McMahon C (2008a) Uncertainty and risk reduction in engineering design embodiment processes. In: Proceedings of the 10th International Design Conference (DESIGN 2008), Dubrovnik, Croatia, pp 143-156

Grebici K, Wynn DC, Clarkson PJ (2008b) Modelling the relationship between uncertainty levels in design descriptions and design process duration. In: Proceedings of the International Conference on Integrated Design and Manufacturing in Mechanical Engineering (IDMME 2008), Beijing, China

Jarrett J, Clarkson PJ (2002) The surge-stagnate model for complex design. Journal of Engineering Design 13(3): 189-196

Johansson C, Parida V, Larsson AC (2009) How are knowledge and information evaluated? - Decision-making in stage-gate processes. In: Proceedings of the 17th International Conference on Engineering Design (ICED'09), Stanford, CA, US

Johnsson S, Malvius D, Bergendahl MN (2009) Performance evaluation of complex product development. In: Proceedings of the 17th International Conference on Engineering Design (ICED'09), Stanford, CA, US

Kellner MI, Madachy RJ, Raffo DM (1999) Software process simulation modeling: Why? What? How? The Journal of Systems and Software 46(2-3): 91-105

Kirk GE (2003) The design of the Rolls-Royce Trent 500 aeroengine. In: Proceedings of the 14th International Conference on Engineering Design (ICED'03), Stockholm, Sweden

Lévárdy V, Browning TR (2009) An adaptive process model to support product development project management. IEEE Transactions on Engineering Management 56(4): 600-620

Neely A (2005) The evolution of performance measurement: Developments in last decade and a research agenda for the next. International Journal of Operations & Production Management 25(12): 1264-1277

O'Donnell FJ, Duffy AHB (2005) Design Performance. Springer, London, UK

Pahl G, Beitz W (1996) Engineering design: A systematic approach. Springer, London, UK

Paquin JP, Couillard J, Ferrard DJ (2000) Assessing and controlling the quality of a project end product: The Earned Quality Method. IEEE Transactions on Engineering Management 47(1): 88-97

Schabacker M (2002) Benefit evaluation of new technologies. In: Proceedings of the ASME International Design Engineering Technical Conference (IDETC/CIE2002), Montreal, Canada

Wynn DC, Eckert CM, Clarkson PJ (2006) Applied signposting: A modeling framework to support design process improvement. In: Proceedings of the ASME International Design Engineering Technical Conferences (IDETC/CIE2006), Philadelphia, PA, US

Zhao Y, Tang LCM, Darlington MJ, Austin SA, Culley SJ (2008) High value information in engineering organisations. International Journal of Information Management 28(4): 246-258

Chapter 14

Continuous Improvement of Mechatronic Product Development Processes

R. Woll, C. Kind and R. Stark

14.1 Introduction

In this paper a methodology for improving mechatronic development processes is presented. The methodology has been developed in the German joint research project MIKADO funded by the German Federal Ministry of Education and Research (BMBF). It describes a comprehensive procedure for process improvement based on a set of reference process models and supported by specific software applications. In a first step the procedure specifies how to generate process models based on an existing process or for planning a new one. In the following analysis phase optimisation potentials in the generated process models are identified. This phase is supported by a set of predefined reference process models that indicate - from a mechatronic view point - at which time during the development and how multidisciplinary coordination activities have to be carried out. A comparison of the generated process models with these reference process models and the application of questionnaires reveals the optimisation potentials that are subsequently used as a basis for improving the generated process model. The improved process model is then used for process control in a workflow management system (WfMS). The methodology has been evaluated against industrial requirements by two medium sized enterprises in the micro-production domain and in the solar energy systems domain.

In Section 14.2 the problem statement will be presented. In Section 14.3 the current state of the art in model-based process improvement approaches is presented and the need for the development of new approaches is motivated. Section 14.4 describes the improvement process, the reference process models and the results of the evaluation. Finally Section 14.5 presents the conclusion drawn from the evaluation and future research topics.

14.2 Challenges in Mechatronic Product Development

A mechatronic product is a product that features mechanical, electronic and software components. During a mechatronic product development project, mechanical, electronic and software engineers have to develop and integrate these components. The development work is devised and assigned to them. In order to integrate these components engineers have to coordinate their development activities and specify the interfaces between the different components. Thus, multidisciplinary coordination activities are required for the successful completion of mechatronical product development projects (Hegewald *et al.,* 2008).

Many mechatronic development projects fail or suffer from unexpected delays or exceeded cost limits because these multidisciplinary coordination activities were not correctly planned and executed. Reasons for insufficient planning or incorrect execution of such activities comprise:

- the planning of mechatronic product development projects and their processes by single experts or teams that are missing representatives from all engineering disciplines;
- the missing focus on coordination activities during process modelling;
- homogenous development teams with engineers from only one discipline that lack exposure to other disciplines work practice.

This can result in:

- misunderstandings between experts from the different disciplines;
- ignorance of the mutual requirements;
- ignorance of the development processes of the other engineering disciplines and of interdependencies between the development processes (Stetter, 2009).

While basic approaches to managing product development processes such as overlapping development phases (Takeuchi and Nonaka, 1986) are common for non-mechatronic products sequential development phases, often starting with mechanical components, followed by electronic and finally software components, are common practice in mechatronic product development.

Studies with multidisciplinary development teams have shown that successful coordination depends primarily on cross-functional knowledge which is missing in 50% of all cases (Boucher and Houlihan, 2008).

Modern virtual product creation provides approaches such as agile process execution or context appropriate information management that offer unique capabilities which have not yet been explored or deployed in mechatronical product development (Stark *et al.,* 2009).

The methodology presented in this paper shall tackle these challenges and explore the use of existing approaches from virtual product creation in order to provide a new approach for the improvement of mechatronic product development processes and the execution of multidisciplinary coordination activities.

14.3 Model-based Process Improvement

The processes in a mechatronic product development project are specific business processes. A business process can be defined as

> *'A set of one or more linked procedures or activities which collectively realise a business objective or policy goal, normally within the context of an organisational structure defining functional roles and relationships.'*
>
> (Workflow Management Coalition, 1999)

The processes in a mechatronic product development project are business processes with the objective of realising a mechatronic product. As the approaches presented in the following apply to a wide range of processes, not only business processes, only the more generic term 'process' will be used.

Process improvement is the activity of ensuring a processes effectiveness and improving its efficiency. Process effectiveness can be measured by the quality of the output, in this case: a mechatronic product. A process is effective if the product matches its specifications and meets customer requirements. Process efficiency is measured by factors such as duration, cost and robustness. Before a process can be improved it has to be surveyed. The result of a process survey is a process model. A process survey is conducted by a process analyst and comprises interviews and workshops with people involved in the surveyed processes. These should be selected from different hierarchies, departments and disciplines. In interviews a process analyst interviews people separately in order to understand their subjective views on a process and identify ambiguous statements. Workshops are conducted in groups where people discuss their different views and build a common understanding of the surveyed process. During a process survey the process analyst documents the process. Sometimes the people participating in a process survey also participate in the activity of documenting the process. The result of the process documentation is a process model. At the end of a process survey the process model has to be approved by the management of the company in which the process survey has been conducted.

A process model is a document that describes a process. It can have different purposes. It can be used for process analysis, for process training, for process documentation or as a guideline for process execution. It describes which activities are happening in which order, who is responsible for executing them, which documents are read, created and edited and which events and decisions occur during the process.

There are graphical and textual representations of a process model. Usually both representations are combined in order to document a process. The typical graphical representation of a process is a process diagram. It visualises a process model for a human observer. Textual representations comprise text in natural language and machine code. Text in natural language is usually used in conjunction with a process diagram and explains the elements in the process diagram in more detail. Computer-readible text is used for saving process diagrams or for process execution in a workflow management system (WfMS).

The representation used to model a process is a process modelling notation. Important graphical, diagram-oriented process modelling notations are the Business Process Modelling Notation (BPMN) (Silver, 2009), Event-Driven Process Chains (EPC) (van der Aaalst, 2006) and the Unified Modelling Language (UML) Activity Diagrams (Booch *et al.*, 2006). These approaches are mainly used for presenting process models to human viewers. The most used textual representation formats for process models are the Business Process Execution Language (BPEL) (Barreto *et al.*, 2007) and the XML Process Definition Language (XPDL) (Shapiro, 2008). These formats define a syntax for machine code and can thus be processed by a WfMS.

Process models that are executable by a WfMS are called workflow models. A workflow model is a process model that specifies the execution characteristics of a process in a WfMS. These characteristics comprise specifications of task descriptions, task assignment policies, databases for saving process data, automated decisions and event listeners. A process model can be extended by these specifications in order to become a workflow model.

The approaches Business Process Reengineering (BPR) (Johansson H *et al.*, 1993) and Business Process Management (BPM) (Ferguson and Stockton, 2006) are approaches for process improvement that rely on the use of process models. The advantage of using process models for process improvement is that it is much easier to identify room for improvement in the composition of activities and regarding resource allocation. BPR defines a set of procedures for reassessing core business processes within a company and redesigning them from scratch. During the process assessment a company's missions, its human resources architecture and its IT architecture are analysed, as well. During the redesign process models that are used as instructions for practice are created and have to be aligned with the company's missions, human resources architecture and IT architecture. BPR does not specify how software applications are used for process improvement. BPM, on the other hand, focuses on the software applications that can be used for process improvement. While there is no proper definition of BPM there are a lot of software applications that are called Business Process Management Systems (BPMS). Thus, we define the BPM approach by the commonalities in the approaches of different BPMS.

Different BPMS implement different procedures for improving processes. Hence, there is not one specific procedure for process improvement that can be associated with the term BPM. Some BPMS feature process simulation functionality. Process simulation can be used in order to support process planning (Wynn and Clarkson, 2009). A common factor of all BPMS is the use of process modelling and process analysis software in combination with a WfMS and a common process modelling notation, the BPMN. Thus, we define BPM as an approach for the technological implementation of process improvement procedures. All procedures presented in this context focus on the improvement of process models and the improvement of processes through workflow execution.

BPR and BPM are rarely used for the improvement of product development processes. The approach of redesigning processes from scratch is not applicable to product development processes because they carry a lot of implicit knowledge. A complete redesign of such processes involves the risk of losing important tacit knowledge. Nevertheless, the methodology postulated by the BPR approach for embedding process models into an organisational context seems very promising for improving product development processes.

The strong focus of BPM on workflow execution is a limiting factor for improving product development processes. Product development processes feature many ad hoc processes and parallel processes that have to be carried out in a flexible manner. WfMS usually constrain the flow of a process to the flow that is specified in a workflow model. Nevertheless, WfMS could be used for the execution of subprocesses within the product development processes that always have to be carried out in a predefined manner. Thus, BPM is an approach that is promising for the improvement of a subset of the product development processes.

14.4 A Methodology for the Continuous Improvement of Process Models

In the research project MIKADO a methodology for the improvement of mechatronic product development processes has been developed that combines components from the approaches BPR and BPM. The methodology postulates that a process can be improved by improving and using a corresponding process model. Thus, process models are used for describing the as-is and the to-be processes and they are used for process analysis. The methodology consists of a procedure for process improvement based on a set of reference process models and it is supported by software applications. In the following the procedure and the reference processes will be described in separate sections. Finally the evaluation of the methodology is presented.

14.4.1 Procedure for Process Improvement

The procedure for process improvement defined in the presented methodology consists of four phases that are depicted in the following figure.

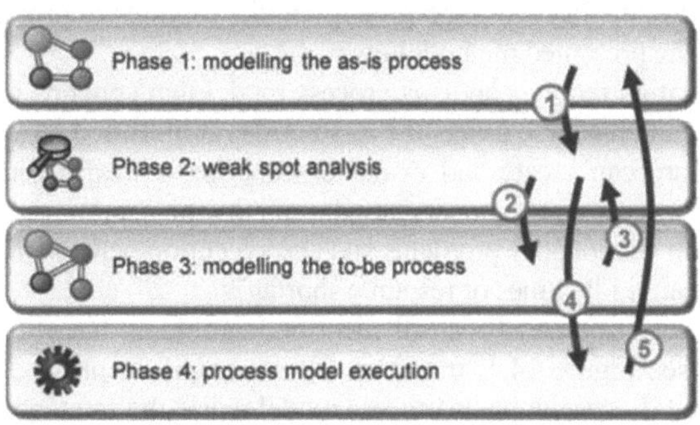

Figure 14.1. The 4 phases of process model improvement

The procedure begins with phase 1, modelling of the as-is process. If the process has not been executed previously then one proceeds with phase 3. If the process has been executed previously but it has not been modelled then a new process model has to be created in phase 1. If the process has already been modelled and executed then the old to-be process model has to be compared with the as-is process, adapted and become the new as-is process model.

In the presented methodology process models are defined using the BPMN. A process model specifies the following process elements:

- all resources required to run a process and their availability;
- all activities in a process, their duration and the resources assigned to them;
- all roles participating in a process and their assignments to activities they are supposed to execute;
- all pieces of data read, edited and created during a process.

BPMN visualises the responsibilities for activities as swimlanes. This makes it easy to identify at which points in a process different disciplines interact. The naming of elements in a BPMN process diagram does not suffice for assuring a common understanding by all readers. Thus, all resources, activities, roles and pieces of data have to be described in a separate text file.

Phase 2, the weak spot analysis, begins with the completion of the as-is process model (see Figure 14.1, transition 1). The weak spot analysis comprises the three steps of process model comparison, process benchmark with reference to process model (see following section) and process simulation that will be described in the following.

The process comparison is carried out in two separate steps. The first step requires an existing to-be process model and the as-is process model. Differences between the existing to-be process model and the as-is process model are identified. This is possible because every deviation from the to-be process model that occurs during phase 4 has to be documented and justified. Deviations comprise omitting predetermined activities, adding new activities, pausing the process and differences in duration, cost and resource usage of single activities. They indicate that either the to-be process model did not describe the real process correctly (case 1) or that people were not sufficiently trained or motivated to use the to-be process model as a guideline (case 2). The required documentation of the deviations allows for distinguishing between these two cases. Case 1 indicates that the to-be process model requires improvement. Case 2 is not addressed in the presented methodology.

Process simulation requires an as-is process model that contains information about the duration, cost and resource usage of every activity in it. Based on this data a process simulation software can create and evaluate scenarios with different values for the duration and resource usage of every single activity in a process (Biantoro, 2007). Process simulation can identify problems such as execution paths in a process that wait for an input and cause idle times or resource shortages.

During the weak spot analysis, all identified problems have to be documented before phase 3 (see Figure 14.1, transition 2) can begin. In phase 3 a to-be process model is modelled. In case no as-is process model exists the reference process models (see following section) can be used for creating a new to-be process model from scratch. In case an as-is process model already exists it is improved and becomes the new to-be process model.

The improvement of an as-is process model comprises changes such as new alignment or ordering of activities, removing or adding of activities, changes in the resource allocation and different values for the expected duration of activities.

After phase 3 phase 2 is repeated and the new to-be process model is searched for weak spots (see Figure 14.1, transition 3). If all weak spots have been resolved one can proceed to the last phase (see Figure 14.1, transition 4). If the weak spot analysis still reveals problems then phase 3 is repeated (see Figure 14.1, transition 2).

In phase 4, the process is executed. A WfMS is used in order to ensure that the process is executed the way it has been specified in the to-be process model. The WfMS routes the process, assigns activities to people and monitors the processes progress. The to-be process model has to be transformed into a workflow model in order to be executable by the WfMS. In addition, the following information is required:

- which people in the company can play which roles defined in the process model and who of them can be assigned which activities;
- how is each single assignment to be presented to its assignee;
- where is the data to be retrieved from and saved to;
- by which rules shall the WfMS chose an assignee for a task if there are multiple people with the same role (load-balancing, round robin, *etc.*).

During process execution the WfMS will:

- remind people of what to do in which order;
- serve as a knowledge carrier structure and provide hints, documentation and check lists along with assigned activities;
- automatically log performance data;
- automatically retrieve and save data from and in the right places;
- notify people when work results they are waiting for have been completed by someone else and provide links to the corresponding digital documents;
- let everybody involved in the process easily track its progress.

The WfMS can help facilitating the exchange of work results between the different disciplines. Engineers from one discipline can check the work progress of the other disciplines and are notified when their work results are completed. The WfMS can also remind engineers that they have to test their work results against the requirements that have been specified by the other engineering disciplines. It can also directly call test routines of other software applications and thus automate testing procedures.

When the process has ended the procedure for improving the process model can be repeated and phase 1 can be started again (see Figure 14.1, transition 5).

14.4.2 Reference Process Models

During the research project MIKADO a set of reference process models for the development of mechatronic products have been developed. The process models have been designed by a German engineering consulting company that participated in the project and that focuses mainly on mechatronic engineering. Thus, the process models reflect expert knowledge from consulting practice and represent best practices.

The reference process models correspond to processes that occur in almost every mechatronic product development project and cover subjects such as requirements management, change management, project management and product design. The reference process models can be used during the presented procedure for process improvement. They can be used as templates for defining new to-be processes in phase 3 or they can be used for evaluating as-is processes in phase 2.

Each reference process model is a set of the following four documents:

- a process diagram;
- a textual description of the elements presented in the diagram;
- a process definition file;
- a questionnaire.

The process diagram depicts the process flow and all activities that have to be carried out during a process. The process diagram is using the Business Process Modelling Language 1.0 (BPMN) as modelling notation. The main goal of the process diagram is to highlight at which points of time during the processes multidisciplinary coordination activities have to be carried out. The process in between these coordination activities is modelled in low detail. That means that blocks of activities such as creating the CAD model of one specific product component are represented as only one activity. Coordination activities, on the other hand, are modelled in high detail. That means that each single activity such as sending a document to a colleague or arranging a meeting for a discussion is modelled as a separate activity.

The textual description provides details for activities, conditions for decisions, roles, data and events in the BPMN process diagram.

The process definition file provides a textual representation of the process diagram in a machine-readable syntax. It uses the XML Process Definition Language (XPDL) data format and can thus be imported into different BPM software applications. The process model described in the process definition file does not describe the processes characteristics for workflow execution. It has to be extended, as described in phase 4 of the procedure, for process improvement in order to become a workflow model.

The process diagram, its textual description and the XPDL file can be used together as a template for instantiating to-be process models in phase 3 of the improvement process. The following steps have to be performed in order to create a specific to-be process model out of the process model defined in the reference process model:

- the roles in the reference process model have to be adapted to the roles in the organisation that is running the development project;
- the passages between the coordination activities have to be detailed;
- the data defined in the reference process model has to be adapted to the kind of documents used in the organisation that is running the development project;
- the resources, durations and priorities assigned to the activities in the process model have to be specified.

The reference process models can also be used in phase 2 of the presented procedure for process improvement. The questionnaire can be used for benchmarking an as-is process model. It is used in interviews. An interviewer addresses the questions in the questionnaire to a process expert and they discuss the different answers. The process expert chooses the answer.

The questionnaire consists of questions regarding the four categories:

- formalisation;
- automation;
- institutionalisation;
- multidisciplinarity.

For each category one question and four possible answers are specified. There are four maturity levels (1 to 4) that a process can be assigned where maturity level 4 is the highest level that can be achieved. The four different answers to each question correspond to the different maturity levels. Thus, the maturity level is evaluated separately for each category. Answering all four questions for the four different categories specified in the questionnaire allows for determining the overall maturity level of the evaluated process. Figure 14.2 shows an extract of the questionnaire of the reference process model for the mechatronic requirements specification process.

Level of Multidisciplinarity	**How are requirements of one engineering discipline communicated to the other disciplines and how are they integrated with the requirements of the other disciplines?**
	1 = Each department or development team creates its own requirements specification. An exchange of requirements between departments or teams does not happen or requirements are communicated informally.
	2 = A dedicated person actively coordinates the exchange of requirements between the departments and teams.
	3 = The integration of mutual requirements is carried out in multidisciplinary review meetings or video conferences.
	4 = Every change in the requirements for one component is automatically tested for consistency with the requirements of other departments or teams.

Figure 14.2. Example: question for the category multidisciplinarity (screenshot)

The category formalisation focuses on the structuring, the documentation and the management of the deliverables and the intermediate data of a process. Questions such as 'Are there templates for deliverables?' have to be answered. Is there a predefined procedure for storing and distributing deliverables? Are there quality checks?

The category automation focuses on the use of software for automating specific tasks in a process. Software such as a Workflow Management System (WfMS) can help automating IT tasks such as automatically sending deliverables to interested receivers upon change or automatically storing them upon completion. Automation also includes functionality such as syntax checks for machine code or software integration test routines.

The category institutionalisation focuses role of the process model in the processes execution and the conformance of the real process with its to-be process model. Process models have to take into account organisational constraints such as the employees work load, qualifications *etc.*. Employees have to know the process models, understand their importance and follow them.

The category multidisciplinarity focuses the quality and degree of the coordination between the stakeholders from the different engineering disciplines. A high level of multidisciplinarity implies the definition of well-defined interfaces for work results such as electronic and software components, frequent communication, collaborative decision making and a strong integration of external development partners.

14.4.3 Evaluation and Results

The improvement process has been evaluated in two small enterprises in the fields of micro-production and solar energy systems. In each company a specific process from a running development project for a mechatronic product was chosen and the presented methodology was applied. In one company the requirements specification process was chosen and in the other company the initial design process was chosen. One of the two companies had existing to-be process models for compliance purposes while the other company did not use process models at all. Process experts performed the first three phases. The chosen processes were already running or had already been completed when the evaluation began. Thus, the new to-be process models from phase 4 could not be evaluated in practice. The process experts created a workflow model that was simulated afterwards. The questionnaires in the reference process models were used during phase 2. The overall evaluation lasted two days in both cases.

The evaluation yielded the following results:

- process models did not play an important role in the companies;
- the as-is process models the process experts created during phase 1 were missing a clearly defined process flow for the execution of multidisciplinary coordination activities; this was also the main weak spot that was identified during phase 2;
- the points in the process where multidisciplinary coordination activities were suggested in the reference process models were adapted by the process experts in both to-be process models;
- the questions in the questionnaire revealed that in both companies processes were not executed as defined in their to-be process models;
- the functionality of the WfMS for routing and monitoring the processes was much appreciated;
- the interviewed process experts criticised that the specification of workflow models was cumbersome;

- the interviewed process experts pointed out that they do not want to use a WfMS for governing flexible subprocesses within a mechatronic product development project.

14.5 Experiences and Future Research

The evaluation of the presented methodology in two companies seems to corroborate the hypothesis that has been postulated in Section 14.2: most process models do not focus on coordination activities in detail. In both companies the procedures for multidisciplinary coordination were neither modelled nor planned. It was not clear at which points during development processes multidisciplinary coordination activities have to be carried out and how they have to be carried out. It has to be pointed out that a statistically valid conclusion required a higher number of companies for the evaluation and the inclusion of medium- and large-sized companies, as well. It also has to be mentioned that no valid conclusion on the duration of the presented procedure can be made because only the first 3 phases have been evaluated and because the duration depends on the use case.

The evaluation also showed that the presented methodology successfully supported process experts to identify when and how multidisciplinary coordination activities have to be carried out during the chosen mechatronic product development processes. Thus, the main goal of the methodology was reached.

For evaluation purposes the procedure for process improvement has been executed only once. Thus, the usefulness of the presented methodology for continuous process improvement could not be evaluated. This remains an open research question.

The presented procedure focuses on assuring that coordination activities are performed at the right time and that they happen at all. Now assuming this is the case, another interesting question is how the coordination activities can be supported by more than just a predefined process? How can the human factors better be taken into consideration? How can software facilitate the communication between the cooperation partners? The presented procedure provides only one building block for an overall approach for improving multidisciplinary collaboration.

The use of a WfMS proved to be either an opportunity for process improvement but also a challenge. Its functionality for routing and monitoring processes was appreciated as long as it concerned processes that are not flexible. This raises the question how a WfMS can be used for governing flexible processes within a mechatronic product development project? Future research should thus focus on the adaptation of WfMS for the governance of product development process.

14.6 Acknowledgements

The authors are grateful for the support of the German Federal Ministry of Education and Research (BMBF) for funding the research project MIKADO.

14.7 References

Barreto C, Bullard V, Erl T, Evdemon J, Jordan D, Moser S, Stout R *et al.* (2007) Web services business process execution language version 2.0 primer. OASIS Web Services Business Process Execution Language (WSBPEL) TC, OASIS Open

Biantoro C (2007) Modellierung und analyse verteilter entwicklungsprozesse für mechatronische systeme. Fraunhofer IRB Verlag, Stuttgart, Germany

Booch G, Rumbaugh J, Jacobson I (2006) The unified modeling language user guide, 2nd edn. Addison-Wesley Professional, Ontario, Canada

Boucher M, Houlihan D (2008) System design: New product development for mechatronics. Aberdeen Group, Boston, MA, US

Ferguson DF, Stockton M (2006) Enterprise business process management - architecture, technology and standards. In: Dustdar S (ed.) Business process management. In: Proceedings of the 4th International Conference on Business Process Management (BPM 2006), Vienna, Austria, pp 1-15

Hegewald T, Kausler B, Thamburaj V, Woll R (2008) Mechatronische produktentwicklungsprozesse beherrschen. Zeitschrift für wirtschaftlichen Fabrikbetrieb 103(10): 732-735

Imai M (1986) Kaizen: The key to Japan's competitive success. McGraw-Hill, New York, NY, US

Johansson H, McHugh P, Pendlebury AJ, Wheeler WA (1993) Business process reengineering: Breakpoint strategies for market dominance. Wiley, New York, NY, US

Shapiro R M (2008) XPDL 2.1 integrating process interchange & BPMN. Workflow Management Coalition, Brussels, Belgium

Silver B (2009) BPMN method and style: A levels-based methodology for BPM process modeling and improvement using BPMN 2.0. Cody-Cassidy Press, US

Stark R, Krause F L, Kind C, Rothenburg U, Müller P, Stöckert H (2009) Competing in engineering design - the role of virtual product creation. In: Proceedings of the 19th CIRP Design Conference - Competitive Design, Cranfield, UK

Stetter R (2009) Das mechatronik-defizit. Computer & Automation 09: 32-34

Takeuchi J, Nonaka I (1986) The new new product development game. Harvard Business Review 64: 137-146

Tennant G (2001) Six Sigma: SPC and TQM in manufacturing and services. Ashgate Publishing, Aldershot, UK

van der Aaalst WMP (2006) Formalization and Verification of Event-driven Process Chains. Information and Software Technology 41(1): 639-650

Workflow Management Coalition (1999). Workflow Management Coalition - Terminology & Glossary. Technical report, The Workflow Management Coalition (WfMC), Brussels, Belgium

Wynn D, Clarkson (2009) Design project planning, monitoring and re-planning through process simulation. In: Proceedings of the 17th International Conference on Engineering Design (ICED'09), Stanford, CA, US

Chapter 15

Interactive Visualisation of Development Processes in Mechatronic Engineering

S. Kahl, J. Gausemeier and R. Dumitrescu

15.1 Introduction

The products of mechanical engineering and related industrial sectors, such as the automotive industry, are often based on the close interaction of mechanics, electronic/electronics and software engineering, which is expressed by the term mechatronics. The design of such systems is an interdisciplinary task. Mechanical, electrical, control and software engineers are involved and have to be coordinated. To approach this challenge, we present the design, implementation and preliminary evaluation of an interactive visualisation tool to illustrate and manage complex development processes of mechatronic system. This is realised by means of a zoomable user interface that is rendered on a high resolution projection system (3860 x 2160).

Modern mechanical engineering products are affected by a high degree of information and communication technology. This is aptly expressed by the term "mechatronics". The term refers to the symbiotic cooperation of mechanics, electrics/electronics, control engineering and software engineering in order to improve the behaviour of a technical system. Modern automobiles, machine tools or airplanes are prominent examples of mechatronic systems. The conceivable development of communication and information technology opens up fascinating perspectives, which go far beyond current standards.

The functionality of mechatronic systems leads to an increased design complexity and requires an effective cooperation and communication between developers from different domains throughout the development process. This postulates a fundamental understanding of the whole system as well as of its development. Established design methodologies, *e.g.* the VDI guideline 2206 (2004), laid the foundation to meet these challenges, but need to be essentially extended and supported by domain-spanning methods and tools. In accordance with these methodologies, the development of mechatronic systems can be basically divided into two main phases: the domain-spanning conceptual design and the domain-specific "concretisation" (see Figure 15.1).

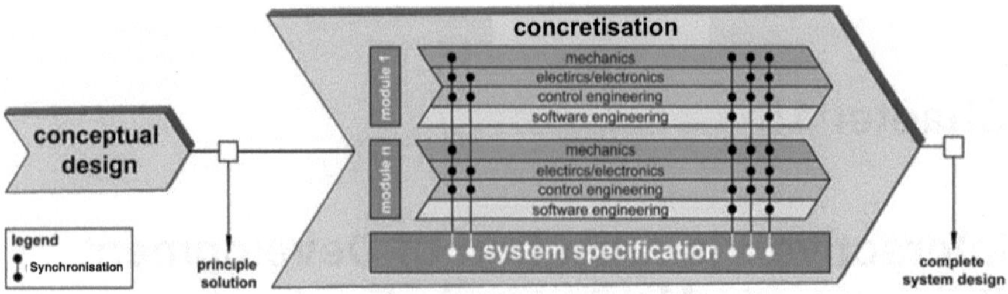

Figure 15.1 Basic Structure of the Development Process

Within the conceptual design, the basic structure and the operation mode of the system are defined and the system is also structured into modules. This division into modules has to result in a development-oriented product structure, which integrates the two basic and mostly contradictious views of shape- and function-oriented structure. One method, for example, is the structuring procedure by Gausemeier *et al.* (2009c). It considers the characteristics and relations of the system's elements as well as changes of the system's behaviour. The result of the process is a development-oriented product structure. The modules defined within this structure can be independently developed, tested, maintained and, if later on necessary, exchanged. All results of the conceptual design are specified in the so-called "principle solution". Based upon the principle solution the subsequent domain-specific "concretisation" is planned and realised. The term "concretisation" describes the domain-specific design of the technical system. The aim of the concretisation phase is the complete description of the system by using the construction structure and the component structure. All defined modules are developed in parallel to each other and each module is developed in parallel within the participating domains. This results in a complex development process with interlaced partial models of the system.

To manage this complexity efficiently, a holistic description of mechatronic systems is required, that regards all engineering domains in an equitable way. Moreover the different development steps, their results and dependences have to be modelled and visualised for the developers in an intuitively understandable way.

Within the Collaborative Research Centre (CRC) 614 "Self-Optimising Systems and Structures in Mechanical Engineering" of the University of Paderborn, a set of specification techniques has been developed in order to describe the system specification of advanced mechatronic systems. By using this set, the system can be described in a holistic domain-spanning way. For further information in respect of the specification technique (please see for example Gausemeier *et al.*, 2009a).

For an intuitively understandable depiction of different development steps, an adequate modelling language is needed. The combination of this modelling language with a matched visualisation has to fulfil the following requirements.

Overview and Details: Developers have to synchronise their work with other activities in the development process and need to identify which process steps relate to others and which can be refined and executed independently. This requires a visual overview of the complete development process as well as a focus on specific and detailed information of individual development tasks and their results.

Search and Filter: The large number of process steps requires an efficient search mechanism. Arbitrary types of process elements should be selectable by filter operations and results should be represented and accessed efficiently.

Interactive Navigation and Presentation: Communication between developers is best realised by means of an interactive visualisation of the complete development process. Users must be able to efficiently navigate through the visualisation and present selected elements at an arbitrary level of detail. Moreover, it should be possible to present a sequence of process steps and their results, for example, to discuss a specific workflow.

15.2 Process Modelling Language OMEGA

For the modelling of complex development processes we use the object-oriented modelling language OMEGA. It has been developed at the Heinz Nixdorf Institute of the University of Paderborn for the modelling and analysis of business processes. A first version was presented in 1995, but since then continuously improved and evaluated within a multitude of industrial projects (Gausemeier *et al.*, 2009b). On the one hand it allows modelling the entire operational structure of an enterprise. On the other hand it is a good instrument for the analysis and planning of business processes because of its intuitively understandable visualisation with an easy and concise imagery.

Modelling the entire operational structure means to depict the objects of the organisational structure as well as the objects of the process organisation in one model. This is realised by assigning the business units to business processes. OMEGA has been validated in several business projects. Experience has shown that business processes, modelled with OMEGA are intuitively to understand. The concentration on the essential objects of a business process makes OMEGA applicable over all enterprise levels. The models are characterised by high flexibility concerning the depiction of project or enterprise specific information.

All essential facts of a business process are illustrated by the imagery of OMEGA. The user is mostly exempt from textual specifications. OMEGA is able to illustrate workflows, information and material flows as well as the parallelism of processes. Figure 15.2 shows the main symbols of the OMEGA constructs. Their semantic is briefly explained in the following.

Business process: A business process is a sequence of activities to create an output or to transform an object. He has a defined starting point (trigger or input) and a defined endpoint (result or output).

Business unit: A business unit represents an element of the organisational structure (person, workplace *etc.*) that executes a business process.

External objects: These are elements of the system environment. Thus they specify the interfaces between the business process and its environment (*e.g.* groups, institutions, enterprises *etc.*). They transmit and receive process objects.

Process objects: Process objects are input and output variables of business processes. Usually is a process object, which has been generated or transformed by a business process, an input object of a subsequent process. For the various types of process objects compare Figure 15.2.

Technical resources: Technical resources support the execution of business processes. OMEGA includes four kinds of technical resources: IT application or storage, equipment, paper storage and material storage.

Communication connection: Communication connections link business processes, external objects and technical resources. Each communication connection has exactly one emitter and one receiver. Placing a business object on a communication connection defines weather it is an information or material flow.

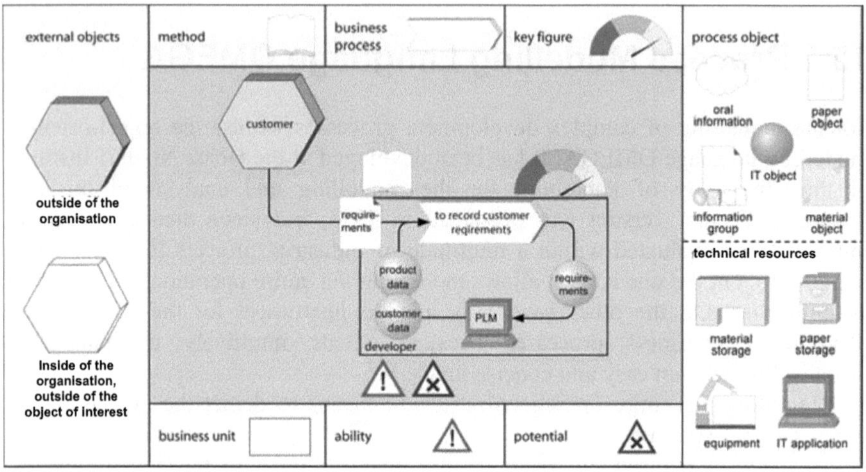

Figure 15.2. Overview of the OMEGA constructs

15.3 Tool Support for the Visualisation

In the following our tool for the visualisation of OMEGA process diagrams will be presented. The aim is to support developers or a complete design team in managing complex development processes of mechatronic systems. With such a tool, we provide a new approach by applying modern visualisation techniques for the entire development process.

15.3.1 Graphic User Interface

Several techniques were implemented to fulfil the requirements mentioned above. The most important is the implementation of the graphical user interface (GUI) as a zoomable user interface (ZUI). Such a GUI allows scaling either several elements or the whole interface (Raskin, 2000). Although a first system based on a zooming interface was developed already at the beginning of the nineties (Perlin *et al.*, 1993), ZUIs became more and more popular in many applications within the last years. Google Maps (http://maps.google.com), Apple iPhone (http://www.apple.com/iphone) or Prezi (http://prezi.com/) are just a brief survey. The basic concept of the GUI of our application is depicted in Figure 15.3.

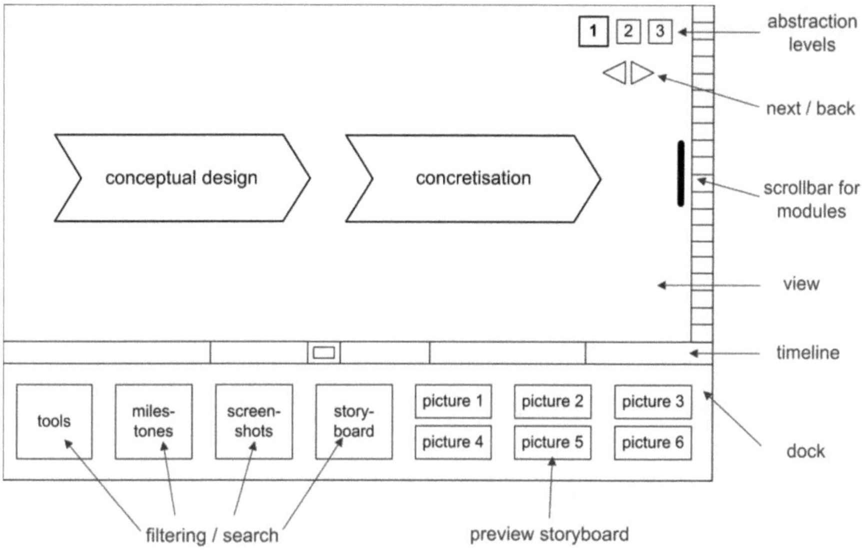

Figure 15.3. Concept of the Graphical User Interface

The ZUI interface of the process visualisation tool enables continuously zooming in and panning over the process model which is displayed in the "view" in the centre of the application. A "dock" at the bottom, which can be switched off, gives an access to various tool functions. To filter and search for certain OMEGA elements the dock provides filter types at the lower left side. In addition a so-called storyboard mode for linear presentations is available. Two coloured scrollbars are implemented and provide an interactive overview: the "timeline" for the horizontal scrolling and a second one for a vertical scrolling between different parallel processes. Those can result from a potential product modularisation. To cope with the complexity of the process diagram during interaction, semantic zooming with three different "abstraction levels" is available. Those can be triggered by buttons on the top of the GUI. At the third abstraction level the complete syntax of OMEGA is getting used. Level two and then level one are aggregating the multitude of processes of the OMEGA diagrams in a semantic correct way in order to avoid an information overload. Furthermore the user always has a fair chance using the abstraction levels to handle process diagrams with 1000 or more process steps, if the aggregation is appropriate, of course. Figure 15.4 shows the final GUI and the three abstraction levels in comparison.

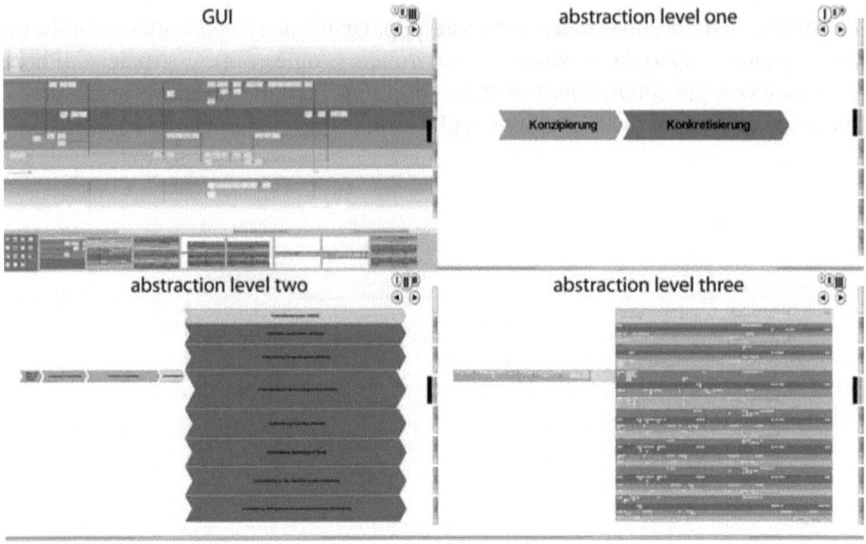

Figure 15.4. Implemented GUI and the three abstraction levels

15.3.2 System Interaction

Besides activating the three abstraction levels at the top right, the user can easily modify the zoom by the input device. Since the application was meant to be used on large screens (See Section 15.4) and standard displays as well, the “Go Pro”-Mouse from Gyration (http://www.gyration.com/index.php/us/products/in-air-micekeyboards/go-pro-air-mouse.html) was chosen as basic input device. It is a cable free gyroscopic device but can also be used as a regular mouse on a table. As a consequence the process visualisation tool works fine with other common mouse devices, too. Additionally to the gyro sensor the “Go Pro”-Mouse is equipped with five buttons facilitating all interacting techniques including ZUI view manipulation, presentation navigation or dock interaction (Geiger *et al.*, 2009a).

The basic mouse interactions and output mapping functions in the view mode are described briefly. A click on a diagram element (*e.g.* a business process) will centre it and focus at the selected element. By drawing a focus rectangle the user can select, focus and zoom in on a group of elements. The mouse wheel realises an intuitive way to zoom in and out, whereas the current position of the mouse pointer indicates the zoom focus. Each interaction that results in a movement is automatically calculated with correct transitions between abstraction levels. Ease-In and Ease-Out at the beginning and end of each animation reduce the cognitive load when following fast movements through the diagram. The dock can be hidden by mouse click to focus only on the ZUI view. Activating one of the filter icons retrieves all positions where the selected icon type exists. It is possible to pan through the results and select one of them. The ZUI view will automatically perform an animated transition to the selected point and adjust the zoom level.

In addition to the mouse interactions there are some keyboard hot keys which are realising auxiliary functions. It is also possible to use multi-touch devices for interaction, which has various benefits in comparison to the mouse/keyboard devices. A preliminary implementation was done on a HP Touchsmart IQ820 from Hewlett-Packard (http://www.hp.com/touchsmart), but needs some further evaluation and optimisation.

15.3.3 Tool Engine

The application was implemented in OpenGL and .net with C# as programming language. The process model was specified in OMEGA with MS Visio and dedicated shape pallets. The engine imports the XML-based Visio files and automatically creates a suitable layout on the screen. It is important that the processes were modelled in Visio correctly by means of the shape pallets and elements connections. The correct appearance, the horizontal layout of the business processes for instance, is realised by the tool engine itself. Moreover, the complete layout is computed in real-time and it is possible to switch to a 3D perspective view.

Typically during the later phases of the development of mechatronic systems distinguishable system modules, like the propulsion drive of an automobile, are getting in terms of concurrent engineering developed in a parallel manner. Reasons for this are to handle the complexity and to accelerate the overall development. The coordination of the workflow and arrangements between the participating developers are getting crucial. We differentiate three types of arrangements: within a system module process, between various system module processes and arrangements according to the overall system design process (see Figure 15.1). Identifying those arrangements is a tough task and essential for the success of the development. Our tool supports this by synchronising automatically those arrangements within the layout of the complete process model. The synchronisation occurs by logical aspects and it is necessary that respective links are set within the process specification in the MS Visio diagrams.

15.3.4 Usage and Presentation Mode

Figure 15.5 gives an idea of how the visual presentation tool looks like in action. For each diagram element it is possible to link a Full-HD resolution image or even a HD movie file. This makes sense especially for the input and output objects of the OMEGA business processes. The user has the opportunity to fade in up to two screenshots at the upper half of the view while still observing the development process or an extract at the lower half. During the overlay of the screenshots, the user still can navigate through the entire process model and switch between arbitrary levels of detail. By clicking on a screenshot, it will close.

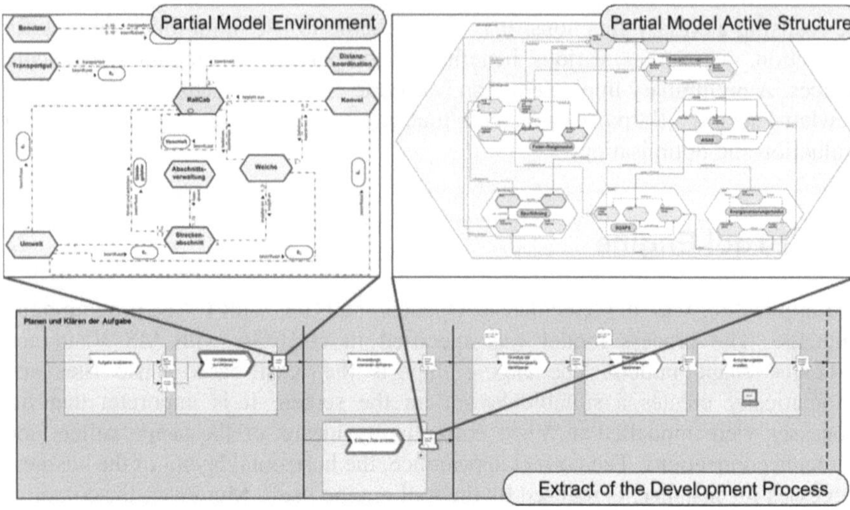

Figure 15.5. Fading in screenshots according to certain process elements

To further support the user in presenting a storyboard mode was applied. It allows creating and viewing a linear tour through the diagram. After activating this mode, the user can navigate to desired positions and add a "screenshot" with position and zoom level parameters to a storyboard and edit in this way a linear list of screenshots. During presentation it is possible to recall those positions from the screenshots only by pressing the forward/backward buttons. Again the presenter is still able to navigate freely through the information space. However, if he wants to proceed with the previously defined storyboard he just presses the respective button and the system calculates an animated transition to the respective screenshot.

15.4 Application and Preliminary Evaluation

15.4.1 Visualisation of the RailCab Development

The process visualisation tool was applied for the development process of an innovative railway prototype called "Neue Bahntechnik Paderborn/RailCab" (http://www-nbp.uni-paderborn.de/) at the University of Paderborn. RailCabs are autonomous vehicles that supply transport for both passengers and cargo on demand without any stops or changes of train and reduce energy consumption by forming convoys. The development process of the RailCab comprises of 850 development steps including about 900 development objects, *e.g.* kinematic models, controller models, state charts, list of requirements, test results *etc.* During the concretisation phase seven system modules, among others a suspension-tilt system and hybrid energy storage system were developed in a parallel way.

The visualisation of the development process of the RailCab prototype runs on projection system, that provides a resolution of 4 x Full HD (3840 x 2160 pixel) (Geiger *et al.*, 2009b). The screen has dimensions of 4.7 m by 2.64 m. The projection system is part of the "High Definition Visualisation Centre for Virtual Prototyping" at the Heinz Nixdorf Institute in Paderborn. Due to the high resolution of the system, developers can focus on OMEGA elements, while, at the same time, do not loose the overview of the whole process model. This allows more efficient design reviews during the product development (see Figure 15.6).

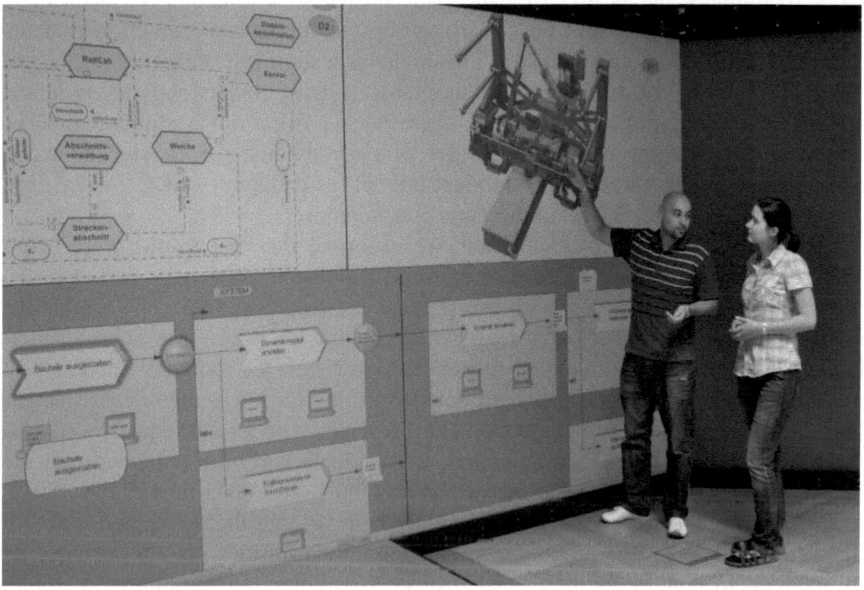

Figure 15.6. Developers in front of the HD process visualisation

The material of the projection screen itself provides a large viewing angle, which ensures a good colour and contrast reproduction even under less optimal viewing angles. So a group of developers is capable of overlooking the whole projection area, while being able to enjoy the full resolution the system provides for displaying graphical items within the ZUI view. This is very important when it comes to reading small characters or detecting small graphical pictograms. There is no need for the user to move closer to or away from the projection screen, in order to have a better view onto the screen.

15.4.2 Preliminary Evaluation

Previous visualisations of the development process of the RailCab have been presented using a large paper sheet or MS PowerPoint. Both techniques were not sufficient according to the requirements from Section 15.1. However, it is necessary to examine, if the process visualisation tool fulfils these requirements, also in comparison to the present methods. Therefore a small experiment was designed to validate our hypothesis that a ZUI application helps users to understand and to handle complex development processes. We defined following test scenario. One presenter explains to a group of 2 to 6 people how the system can be developed simulating a regular design review meeting. This scenario is carried out with three alternative visualisations: a large paper sheet model (3.0 m x 1.5 m) illustrating the complete process model of the RailCab, a PowerPoint presentation that was successfully used during lectures and project meetings, and the process visualisation tool described in previous sections. Each presentation takes 10 minutes and is given by the same presenter. The audience, mechanical engineering students, is split up into six groups. After each presentation students are asked to fill out a questionnaire. This includes a short test with questions about the process model and evaluates if the subjects have understood the technical content of presentation. Additionally, the students filled out a short qualitative review with questions on the suitability of each medium. Each student group participates in all three presentations but in a different order. After each presentation the subjects fill out the same questionnaire.

Although not statistical significant, we observed that test results were better (according to the number of correct answers) if subjects participate earlier in the presentation that was given with the process visualisation tool. The qualitative review showed that participants had some problems to follow the unusual, non-linear presentation with animated transitions. However, the tool was rated as best medium and the amount of provided detailed information was significantly larger than the other presentation variants. The presenter's style was equally rated as "good" in all treatments. Summarising, the evaluation results indicated that the process visualisation tool performed similar to the well-known presentation forms of PowerPoint and paper sheet. We strongly believe, that the developed tool will distinctly overcome the common presentation possibilities, if the amount of information to convey increases. We expect to validate our hypothesis in future experiments.

15.5 Summary and Future Work

The development of mechatronic systems is still a challenge due to its interdisciplinarity. In this contribution we presented a tool for the intuitively understandable modelling and visualisation of complex development processes. We introduced OMEGA as modelling language for our visualisation tool. The first application of our tool is the development process of the RailCab prototype. Its powerfulness was attested by a preliminary evaluation.

Future work will deal with a dynamic, situation-specific adaptation of development processes. Dynamic adaptations can be necessary; because of general conditions, design objectives or the system's architecture might change during the development. All in all no development process is equal to another (Albers *et al.*, 2007). Therefore an automated process management should be implemented. Depending on the general conditions, the actual design objectives and the kind of development object, the development process should be planned, monitored and if necessary adapted.

In distinction to other approaches (*e.g.* Westfechtel, 2001; Krause *et al.*, 2004) we will apply the paradigm of self-optimisation (Adelt *et al.*, 2009) for the automated management of development processes. Thus the actual degree of fulfilment of the development objectives will be analysed and evaluated. Based on this evaluation the parameters or the structure of the current development process is adapted if necessary.

In the first step we focus on the generation of alternative plans for development processes. Based on the behaviour planning techniques for interacting intelligent mechatronic systems in non deterministic environments by Klöpper *et al.* (2009) we develop a two-stage, knowledge based approach. Figure 15.7 shows the main elements and their interconnection.

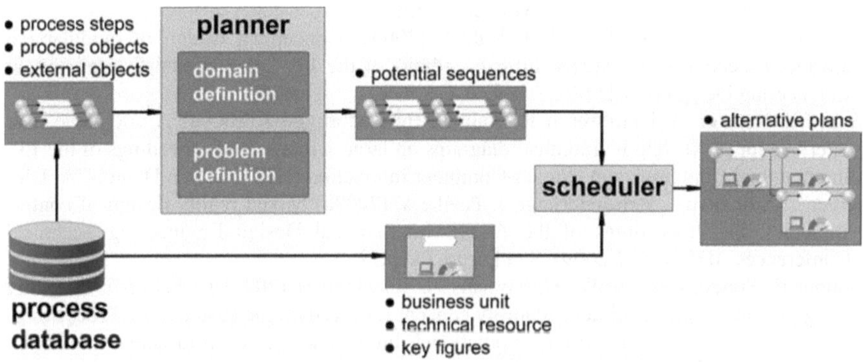

Figure 15.7. Structure of the process planning approach

In the first step we use the HTN-planner JSHOP2 (http://www.cs.umd.edu/projects/shop/index.html) to identify all potential process sequences that lead to the requested result of a development task. For this purpose potential process steps and their input and output objects (process objects, external objects) need to be stored in the process database. In the second step the component scheduler adds additional information (business units, technical resources, key figures) to the potential sequences and selects realisable ones. The described procedure results in a set of alternative plans for the fulfilment of a development task. The identified alternative plans should form the basis for the selection of an initial development process and later on for the automatic adaptation of the development process.

15.6 Acknowledgement

This contribution was developed in the course of the Collaborative Research Centre 614 "Self-Optimising Concepts and Structures in Mechanical Engineering" funded by the German Research Foundation (DFG). One author obtains a research fellowship by the International Graduate School "Dynamic Intelligent Systems" supported by the federal state North Rhine-Westphalia, Germany.

15.7 References

Albers A, Meboldt M (2007) iPeMM. Integrated product development process management model, based on systems engineering and systematic problem solving. In: Proceedings of the 16th International Conference on Engineering Design (ICED'07), Paris, France

Gausemeier J, Frank U, Donoth J, Kahl S (2009a) Specification technique for the description of self-optimizing mechatronic systems. Research in Engineering Design 20(4): 201-224

Gausemeier J, Plass C, Wenzelmann C (2009b) Zukunftsorientierte unternehmensgestaltung: Strategien, geschäftsprozesse und it-systeme für die produktion von morgen. Carl Hanser Verlag, Munich, Germany

Gausemeier J, Steffen D, Donoth J, Kahl S (2009c) Conceptual design of modularized advanced mechatronic systems. In: Proceedings of the 17th International Conference on Engineering Design (ICED'09), Stanford, CA, US

Geiger C, Reckter H, Dumitrescu R, Kahl S, Berssenbrügge J (2009a) A zoomable user interface for presenting hierarchical diagrams on large screens. In: Proceedings of the 13th International Conference on Human-Computer Interaction (HCI'09), San Diego, CA, US

Geiger C, Stöcklein J, Berssenbrügge J, Paelke V (2009b) Mixed reality design of control strategies. In: Proceedings of the ASME International Design Engineering Technical Conferences (IDETC/CIE2009), San Diego, CA, US

Klöpper B, Sondermann-Wölke C, Romaus C, Voecking H (2009) Probabilistic planning integrated in a multi-level dependability concept for mechatronic systems. In: Proceedings of the IEEE Symposium on Computational Intelligence in Control and Automation (CICA'09), Nashville, TN, US, pp 104-111

Krause F-L, Kind C, Voigtberger J (2004) Adaptive modelling and simulation of product development processes. CIRP Annals - Manufacturing Technology 53(1): 135-138

Perlin K, Fox D, (1993) Pad: An alternative approach to the computer interface. In: Proceedings of the 20th Annual Conference on Computer graphics and interactive techniques (SIGGRAPH 93), Anaheim, CA, US

Raskin J (2000) The humane interface: New directions for designing interactive systems. Addison-Wesley, New York, NY, US

VDI-guideline 2206 (2004) Design methodology for mechatronic systems. Beuth-Verlag, Berlin, Germany

Westfechtel B (2001) Ein graphbasiertes managementsystem für dynmaische entwicklungsprozesse. Informatik Forschung und Entwicklung 16(3): 125-144

Chapter 16

Management of a Design System in a Collaborative Design Environment Using PEGASE

V. Robin, C. Merlo, G. Pol and P. Girard

16.1 Introduction

PLM (Product Lifecycle Management) systems are deployed within companies to support product data structuring and management throughout the product development process. They manage information through document management and especially product data evolution using predefined workflows (Liu and Xu, 2001). Actual PLM systems integrate Internet-based technologies and offer groupware-like functionalities (Johansen, 1988; Eynard *et al.*, 2002) for collaboration among actors. Several PLM systems have recently introduced project management functionalities (Saaksvuori and Immoen, 2004). Most of the time these functionalities allow the formalisation of tasks and milestones schedule. Nevertheless this project implementation reveals strong limitations if correlated with design coordination. On the one hand the management of deadlines and the modifications of task sequences can be made dynamically. On the other hand, it is not possible to 'reuse' predefined task sequences or to 'redo' specific ones as compared to workflow capabilities. Main limitation concerns the impossibility to drive documents life cycles from the tasks schedule. Concerning the SMEs, when implementing a PLM system in an SME, we face two antagonistic problems: first to improve the level of formalisation of information flows and second to keep a certain level of flexibility (Weber *et al.*, 2002). To favour flexibility of the design process and coordinate them, formalisation has to consider the definition of the product and its evolution, the objectives of design constrained by the organisation of the company (Mintzberg, 1989) but also the factors that influence the system as technologies, human resources and physical implementations (Wang *et al.*, 2002). As a first but perhaps limited solution the coordination through PLM systems is to be studied in order to integrate document workflows and to introduce flexibility into such workflows (Saikali, 2001) for global project coordination. A second solution is to implement a specific tool for managing design processes in a context of integrated design coordination, as a tool based on the CoMoDe model (Gonnet *et al.*, 2007). We chose this solution and developed a prototype of software dedicated to design coordination: PEGASE. In this paper, we present some results of the IPPOP project and we focus more particularly on the software application PEGASE. The first part of the paper focuses on design coordination and introduces a new approach based on collaboration analysis to increase the level of formalisation of design processes. Then we present the integrated product - process - organisation model of IPPOP project. On the basis of this model we propose then a detailed presentation of PEGASE, a prototype of software supporting actors throughout the design project. We show how the prototype makes it possible to model and to follow-up the evolution of the design system but also to create a project and to follow-up its progress. Finally, we describe perspectives for the development of new functionalities to evolve our prototype of the software.

16.2 Design Coordination in an SME

16.2.1 Design Coordination

Today design projects depend on the ability to coordinate and to control the collaboration between the numerous actors participating in such projects: *e.g.* designers, experts from different disciplines and with different experiences, or external partners. Coordination and control of engineering design are part of a global approach for the development of new products which implies the need to identify the different situations occurring during the design process and the adequate resources to satisfy design objectives. In design project management, the control of the progress of design process can be defined as the understanding of existing design situations (in the real world) in order to evaluate them and take decisions that will modify and improve the future process, according to design objectives given by customer specifications or issued from the company strategy. So, management of design projects is a problem of decision-making to support designers in their activities (Girard and Doumeingts, 2004) in order for them to achieve an objective in a specific context. This specific context could have an influence on the project and refers to the environment of the enterprise (society, market, subcontractors, *etc.*) but also to its organisation (Robin *et al.*, 2007) (Figure 16.1, 1). Influences of the context affect each entity of the organisation and consequently oblige to adopt a multi-level project management adapted to each decision-maker at each decision-level (Figure 16.1, 2) in order to provide to each project manager a set of information representative of the real state of design situation. All the information has to be synchronised for each project in the organisation to ensure coherence of the project management (Figure 16.1, 3). Information has also to be continuously defined and characterised to permit an efficient decision-making during the project progress (Figure 16.1, 4). It is possible only if each information flow, for each project is traced, analysed and exploited to follow-up the progress of the project (Figure 16.1, 5).

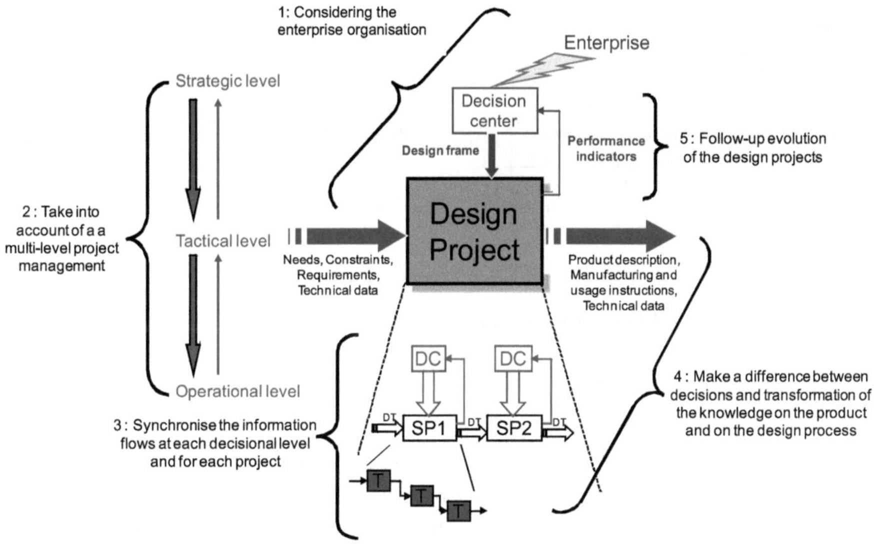

Figure 16.1. Coordination of design activities

In an SME design projects are generally different and require a specific study for each customer's specifications. Most of the time, the small structure of the SME does not ensure project management in a routine way and leads to combine various responsibilities. Indeed there are not enough actors to fulfil each design role, so most of the actors have various design roles in a project. Consequently the role of informal relationships is very important in the SMEs in order that each design stakeholder may help each other without rigid formalities. Thus, the combination of various responsibilities and the informal relationships lead to a high level of workload because informal tasks are added to the official ones. Accordingly SMEs have to manage deadlines by setting an order of priorities on design tasks according to the objectives. Another point specific to SMEs is their project structures with a rigid formalisation of their processes at a macro level and a very flexible non-formalisation of the detailed processes which allow informal relationships into the project. In this context, the project manager coordinates (Figure 16.2) by analysing the requirements from the customer, after which he defines the project team with its internal organisation. He then defines the sub-phase of the project plan and activities in each sub-phase, next he defines a plan to control the project progress and finally he applies this control plan. Periodically he controls project progress and makes the adequate modifications according to the results and the design objectives.

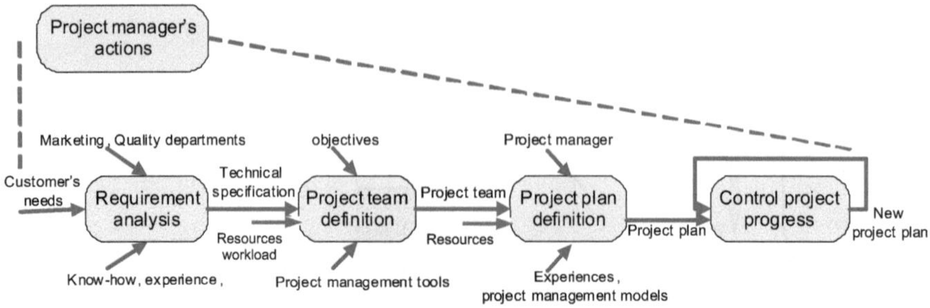

Figure 16.2. Coordination of design activities

For supporting the project manager when defining and controlling the design process, it is necessary to:

- help him to characterise the process to be scheduled;
- help him to manage the design process with a global view, but also all the local processes distributed through the different sub-projects and departments and teams of the company and partners.

In the next section the problem of process characterisation and management is addressed and we try to provide some potential answers for the use of PEGASE, our prototype software tool dedicated to design coordination.

16.3 PEGASE: A Prototype Software to Control Design Projects

16.3.1 Presentation of PEGASE

16.3.1.1 The IPPOP Integrated Product-process-organisation Model

Development and implementation of our prototype of software tool is based on the results obtained during the IPPOP project (Integration of Product, Process and Organisation to improve the Performance in design) (Roucoules *et al.*, 2006). One of the principal objectives was to formalise a model integrating three complementary dimensions: the modelling of the product, the modelling of the design process and the modelling of the organisation where this process proceeds (Robin and Girard, 2006). This model constitutes the core of a collaborative environment making it possible for the actors to control the design process and to be interconnected with the existing applications (XAO tools, PLM systems). In this context a methodology and a prototype of the software supporting the methodology (PEGASE) were developed. The prototype makes it possible:

- to correlate the organisation of the company and the structure of the project;
- to structure the project and to plan the design process by integrating performance indicators concerning product, process and organisation;

- to ensure the follow-up of the project through this integration and these performance indicators.

To implement this model and to allow the effective control of the design projects, the company, its organisation and its resources have to be described. Following section details the development phase of PEGASE based on an analysis of the design process and of the mechanisms of decision-making throughout the product development. A presentation of the prototype of software is also provided.

16.3.2 Control the Design Process Using PEGASE

The detailed analysis of design processes and of the mechanisms of decision-making throughout the product development allows identifying elements that have to be managed to control design process (Robin *et al.*, 2007). PEGASE has been developed to integrate and manage all these elements. The integration phase concerns the implementation and the configuration of the database. It is dedicated to the administrator of the system. To manage evolution of the design process, projects have to be structured, planned and resources have to be allocated. This phase is realised by the project manager. Finally, PEGASE controls project evolution by managing the realisation of the designers' activities and helping managers to follow-up the project. In a nutshell, control of the design process using PEGASE results in several actions from the genesis of the project to its closure:

- implementation and configuration of the database;
- structuring and planning the projects and allocated resources:
    - after the project was initialised and the objectives of the company were specified, the head of project structure his project in order to achieve his goals;
    - he defines several sub-projects for which he specifies the objectives and the persons in charge (as local decision centres);
    - he associates input technical data necessary to the designers to achieve their goals, and output technical data corresponding to the achievement of these objectives;
    - he defines a plan of the activities to be carried out by specifying their technical objectives and their data.
- realise the activities and follow-up the design projects:
    - to allow the follow-up of the project, the designers generate the awaited technical data and evaluate the required performance indicators.

These actions associated with the integrated model ensure that the organisation of the company, the multilevel management of the projects, the differentiation between the decisions and the transformation of product-process knowledge, the synchronisation of informational flows and finally the follow-up of the projects are taken into account.

16.3.2.1 Case Study

The industrial case study has been achieved in the EDERENA Company, an SME which, some years ago, developed a new means of manufacturing structures using honeycomb sub-assemblies. This innovation confers lightness and significant vibration absorption on products whilst maintaining similar rigidity to steel. The company has captured several markets with products manufactured using its technology and consequently the number of employees grew from 4 to 40 over 10 years. Over this period the organisational structure and internal processes have not been formally revised. The objective of our study was to help the company to reorganise and to introduce the role of "design project manager" in order to manage further growth. In this context, problems of organisation, project management and relationships with suppliers, customers, and subcontractors come into play. We have first studied and analysed the company's design and industrialisation department. Then we have formalised: a new organisational structure, the processes of new products development and the management of technical information and of product data. After this first phase we have focused our work on the study of collaboration and relationships between actors and on the design project coordination.

16.3.2.2 From Collaboration Analysis to Detailed Processes Characterisation

After four months of tracking projects in our industrial partner, four different projects have been deeply analysed and more than one hundred collaborative events have been stored. The following example illustrates the consequences of such analyses on the project management: the introduction of flexibility and detailed implementation of design processes. The example is based on the CND (Customer's Need Definition) process which corresponds to the initial financial quotation phase of the design for the customer. Initially, the CND document was managed by the marketing person who builds the document in collaboration with the customer. Indeed this step defines the specification of the product on the basis of the need expressed by the customer. The activities of this phase were (Figure 16.3):

- Definition of the CND document by marketing person with the customer (task A11).
- Validation of the document (task A12).
- Notification that the document is complete (between A12 and A13) to the technical department and that a designer has to make the quotation (future tasks A13).

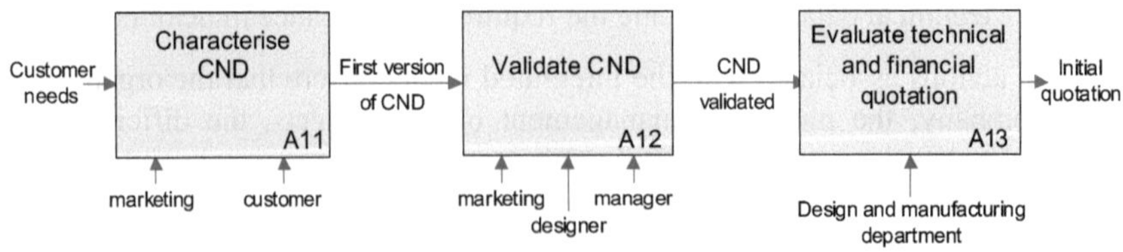

Figure 16.3. Coordination of design activities

The analysis of this initial collaborative situation through several projects allows for identifying that the CND process description incorporates neither details on the way to achieve the tasks, nor flexibility. Moreover the marketing person does not always have the necessary technical skills for all customers, and furthermore he does not have enough time to carry out all the CND processes. So the problem of customer data management appears between the marketing and technical departments. Analysis of the collaboration allows the analyst to define guidelines and more detailed processes. In this way, the CND process is updated with an increased level of granularity based on the guidelines from the collaboration analysis. Consequently a new process is proposed in Figure 16.4 and details previous task A11. The marketing person first evaluates the needs of the customer (task A111), then he can:

- reject directly the customer request, if the customer needs are not appropriated for the company (not formalised);
- make a visit to the customer alone (task A112) before sending the detailed needs to the designer (task A114) or with a designer (task A113);
- or directly send the needs to the designer if they are detailed enough (task A114).

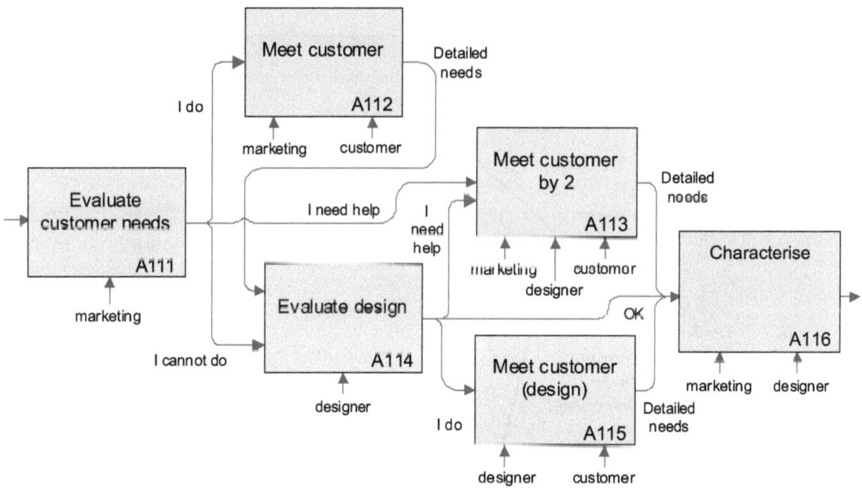

Figure 16.4. Coordination of design activities

Afterwards when the designer evaluates the design (A114), he can meet the customer alone (A115) or with the marketing person (A113), or directly characterise the CND document (A116). At each task marketing person or designer have the possibility to end the process. As a conclusion the project manager has the possibility to automate the design process by implementing a PLM system with this process. The first node of flexibility is the task A11 because the detailed sub-level may not be scheduled for a specific reason. Next nodes of flexibility are associated to tasks A111 and A114 as choices exist for the owner of the task. Next section develops the implementation of such process into a PLM system.

16.3.2.3 Implementation and Configuration of the Database

Within the framework of GRAI R&D approach (Girard and Merlo, 2003), the modelling of a company makes it possible to formalise its organisation (functional decomposition and decisional system) and its technological system (design process). Via an administrator access, the organisation is seized within PEGASE (Figure 16.2). The structure of the decisional system is defined using GRAI R&D grid. Decision centres are identified and their temporal range, their nature and information flows connecting these centres are identified too. This structure is deployed in PEGASE by associating each element of the organisation (plant, department, service) and the corresponding decision centres and by connecting them between specifying information flows (Figure 16.5). So, the administrator configures information flows which will have to be implemented in the course of the project by the various local coordinators implied in order to ensure the coherence of their communication and their decision-makings. The administrator access also permits to define the whole of the resources: human, material and software. Humans' competencies are also managed and are specified according to competencies matrix of the company. Through the management of their competencies, human resources could be selected by the decision-makers to achieve activities during the design projects.

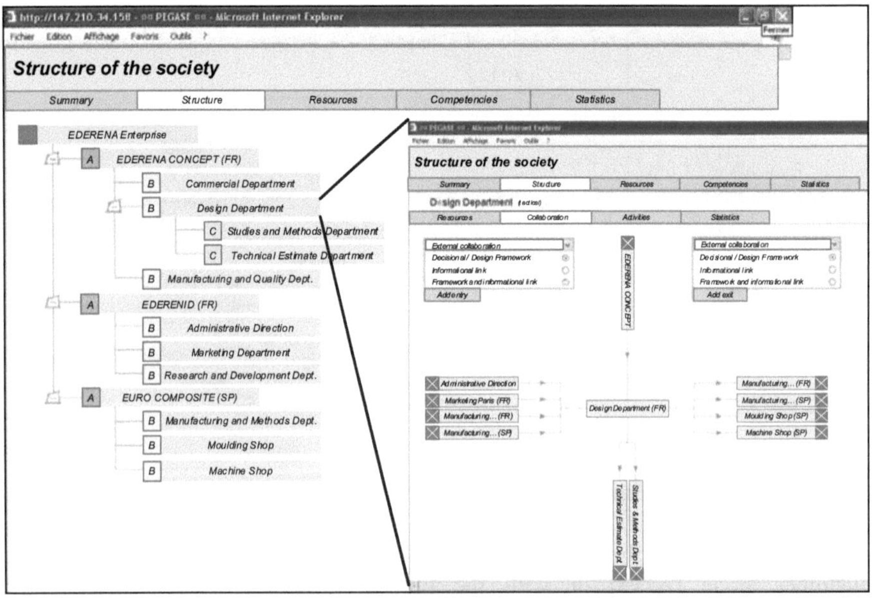

Figure 16.5. Coordination of design activities

16.3.2.4 Structure, Plan and Follow-up a Design Project

When all the structure of the enterprise is defined and implemented in the prototype, the administrator deploys the design process modelled in the organisation by associating to each decision centre the sequences of tasks. This process could be formalised according to the quality procedures of the company. In our case study, we implemented the activities composing process for A11 task (Figure 16.3) that are associated to the design department of the EDERENA Concept Company (Figure 16.6). PEGASE ensures the progress of the sequence of activities and manages the information flows between each one. That is to say that all the information flows through the structure are controlled. The modelling of the decisional and functional structures of the company allows managing processes and collaborations between each entity. The definition of all the processes and of the activities that composed them permits to control progress of each project of an enterprise with a multi-decisional level point of view.

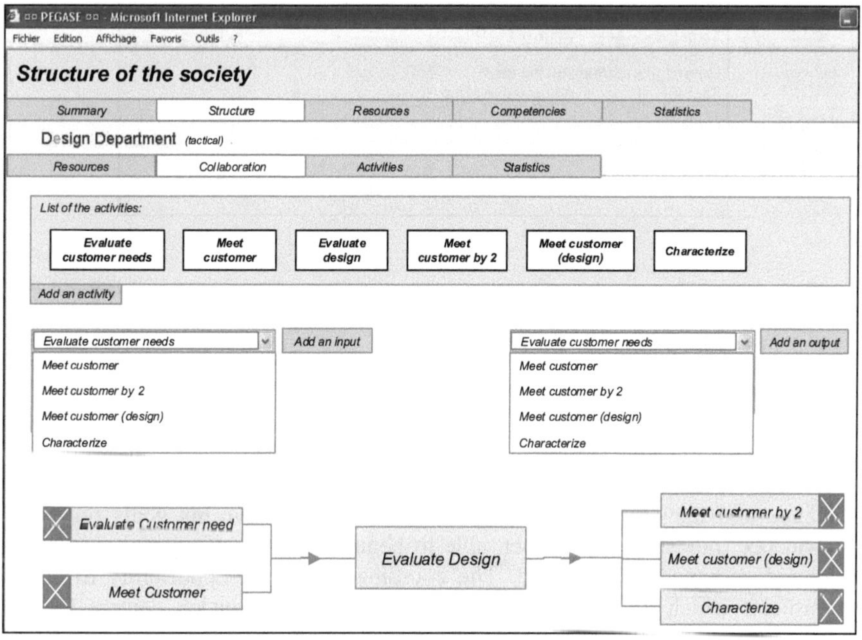

Figure 16.6. Coordination of design activities

When configuration is completed, PEGASE is operational in order to ensure the control of the design. So, the administrator creates and initialises a project by sending the decision frame and associated design frame to the decision centres concerned in the organisation. When the project and the process are initialised, PEGASE systematically informs the users of the new events which relate to them. So each coordinator is informed of his new statute when he is connected. He has information about the organisational structure of the company in order to know the other coordinators with whom collaborations will be established. He is able to reach directly the details of the new project and to reach the decision frame that is sent by the upper decisional level. Figure 16.7 presents a design framework concerning the "evaluate design" activity of the process for the A11 task.

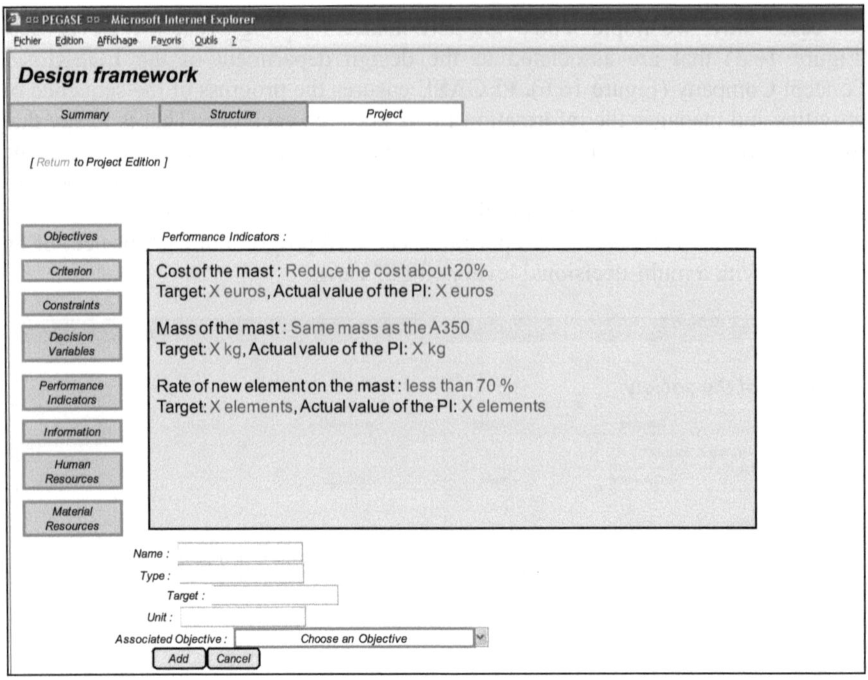

Figure 16.7. Coordination of design activities

The decision frame enables an actor to know his context of work: his objectives, his criterion and decision variables, his constraints, his performance indicators and the resources which are allocated to achieve his goals regarding performance indicators. He is then able to begin the phase of control previously structured, assigned and planned. The coordinator has the opportunity to create sub-projects which will be automatically associated to decision centres for the lower decisional level. He defines finally the tasks to be carried out by completing whole or part of the tasks specified by the administrator, or by introducing new tasks depending on the needs for the project. It guarantees the flexibility of the design process evolution during the project. By using the preset informational links, PEGASE informs each new local coordinator of subprojects and each designer affected to specific tasks of their objectives. Project managers and the designers have the same type of GUI (Figure 16.7) to understand the context in which they must carry out their tasks. The difference is that project managers create performance indicators and designers could just complete these indicators. They must, at the end of their task, indicate the values of the performance indicators. The coordinator is informed of the new values to his next connection. If the indicators don't correspond to the attended values, he analyses the situation and then could decide: to start new activities, to modify some elements of the decision frame (objectives, constraints, resources). At this moment PEGASE allows modelling structure of the enterprise, processes and activities to manage relationships between each entities of the structure. Control of the design project's progress is possible due to the decomposition of the project in the structure and the follow-up of the evolution of the input and the output of the activities. Better the definition of the design activities is, more efficient the design process management is.

16.4 Synthesis and Conclusion

In the worldwide competition among companies, the development of new products has become a challenge where innovation and coordination of design process are two main keys for success. In SMEs design activity is not completely structured and controlled due to the high level of flexibility of processes. At the same time PLM systems help to rationalise basic design processes and are the main information systems managing the product life cycle in companies. In this paper we have focused on the proposal of a framework for design coordination implemented through a PLM system. First we have proposed an adapted method for implementing PLM systems in order to take into account both more detailed process definition and flexibility by using the analysis of collaborative practices. Second we present PEGASE, a prototype of software tool to support design project management. Work reported here constitutes a specific application of the results from IPPOP project in order to make effective the control of the design. The approach suggested is based on an integrated product - process - organisation model and on a methodology modelling the design system and structuring, planning and following-up the design projects. A case study describing experimentation in a SME validates the methodology. PEGASE is a prototype whose first version only takes into account organisation and process dimensions. Influences of each element of the design context are considered since functional and decisional structures of the design system are identified and external environment (subcontractors, competitors, customers, *etc.*) is integrated. PEGASE takes part in the project management from the creation of the project to its closure. All the projects of the company could be managed and controlled through the prototype which provides and capitalises information on the project, on the resources (competencies, availability, *etc.*) to follow-up evolution of the design system. The prototype is perfectible and needs to be sophisticated. Thus, the management of the activities only concerns sequential process. As the first results are significant enough to justify the interest of our multi-level workflow framework for design process management we will implement in PEGASE all the functionalities of this framework. The objective is to use workflow technologies such as those existing within PDM tools to provide a great variety of specification for the managers while providing mechanisms of multi-level synchronisation (Pol *et al.*, 2005). The prototype currently only makes it possible to define indicators manually and it is envisaged thereafter to enrich this functionality by automatisms that permit to identify indicators along the evolution of the process, the product and the organisation. Moreover, the taking into account of the material resources for their management is not effective yet and the integration of the CO^2MED tool (Rose *et al.*, 2007) - a tool for management of collaborations between actors - is also under study.

16.5 References

Eynard B, Merlo C, Carratt B (2002) Aeronautics product development and certification workflow based on process modelling. In: Proceedings of the 8th International Conference on Concurrent Enterprising (ICE 2002), Rome, Italy

Girard P, Merlo C (2003) GRAI-engineering methodology for design performance improvement. In: Proceedings of the 14th International Conference on Engineering Design (ICED'03), Stockholm, Sweden

Girard P, Doumeingts G (2004) Modelling of the engineering design system to improve performance. Computers & Industrial Engineering, 46(1): 43-67

Gonnet S, Henning G, Leone H (2007) A model for capturing and representing the engineering design process. Expert Systems with Applications, 33: 881-902

Johansen R (1988) Groupware: computer support for business teams. The Free Press, New York, NY, US

Liu DT, Xu XW (2001) A review of web-based product data management systems. Computers in Industry, 44: 251-262

Mintzberg H (1989) Le management: voyage au centre des organisations. Les Editions d'Organisation, Paris, France

Pol G, Merlo C, Jared G, Legardeur J (2005) From PDM systems to integrated project management systems: a case study. In: Proceedings of the International conference on Product Lifecycle Management (PLM'05), Lyon, France

Robin V, Girard P (2006) An integrated product-process-organisation model to manage design system. In: Proceedings of the 4th CESA Multiconference on "Computational Engineering in Systems Applications" (CESA 2006), Beijing, China

Robin V, Rose B, Girard P (2007) Modelling collaborative knowledge to support engineering design project manager. Computers in Industry, 58(2): 188-198

Rose B, Robin V, Girard P, Lombard M (2007) Management of engineering design process in collaborative situation. International Journal of Product Lifecycle Management, 2(1): 84-103

Roucoules L, Noel F, Teissandier D, Lombard M, Debarbouillé G, Girard P *et al.* (2006) IPPOP: an opensource collaborative design platform to link product, design process and industrial organisation information. In: Proceedings of the International Conference on Integrated Design and Manufacturing in Mechanical Engineering (IDMME 2006), Grenoble, France

Saaksvuori A, Immoen A (2004) Product lifecycle management. Springer-Verlag, Berlin, Germany

Saikali K (2001) Flexibilité des workflows par l'approche objet: 2Flow: un framework pour workflows flexibles. PhD Thesis, Ecole Centrale de Lyon, Lyon, France

Wang F, Mills JJ, Devarajan V (2002) A conceptual approach managing design resource. Computers in Industry, 47: 169-183

Weber C, Werner H, Deubel T (2002) A different view on PDM and its future potentials. In: Proceedings of the 7th International Design Conference (DESIGN 2002), Dubrovnik, Croatia

Chapter 17

Towards a Robust Process for Integrating Innovations into Vehicle Projects

G. Buet, T. Gidel and D. Millet

17.1 Context

17.1.1 Characteristics of a Robust "Touch-down" Process

No matter which car maker is observed, it appears that very few innovations actually find their way into vehicle development projects, compared to the number of ideas originally imagined.

Although it is normal practice to filter out many innovations, it is essential to maintain a certain number within the vehicle development projects. Otherwise there is a risk of not being able to keep up with market expectations or of being out of step with the competitors' market offerings.

This difficulty of transforming good ideas into innovations that find their place in vehicle development projects may be attributed to the difficulty in converging innovation development with vehicle development, a process that we shall refer to in the rest of this article as the "touch-down" process (Buet *et al.*, 2008).

This term stems from an analogy that may be made with an aircraft (innovation projects) landing on an aircraft-carrier (vehicle development projects). While landing, it is essential to specify all the conditions required, to apply all defined processes, but also to know how to react to events in order to make a successful "touch-down".

In order to keep abreast of the market, motor manufacturers are faced with co-ordinating innovations that have not always achieved a sufficient degree of maturity with the vehicle development projects that are likely to be their platform. The notion of "touch-down" is typified by the integration such innovations in vehicle development projects.

> *'The integration process itself is proving problematical in that new technology fields, organizations and timescales differ considerably from those applicable to vehicle development projects. [...] It is a complex process to successfully converge*

> *new technology developments (available at the right moment and at the right level of maturity) with product development projects.*'
>
> (Buet *et al.*, 2008)

The innovation "touch-down" process can be divided into several phases - innovation genesis, selection and, finally, lock-on to the vehicle development project.

The first phase includes identifying and selecting innovations that would appear to be relevant to one or more selected vehicle projects. This phase comes to an end with the decision to integrate one or more innovations into one or several vehicle development projects. This decision, which is taken sufficiently up-stream in the development process so as to anticipate the risk factor, may be challenged, which is not the case once it has been decided to "lock-on". Lock-on denotes the notion of effectively mating innovations with the vehicle development project. In our aviation analogy, it would be the hitching of the aircraft to the aircraft carrier with its tail-hook.

The decision to lock on may be expressed as the risk of integrating versus the risk of not integrating a given innovation. When the risk of not integrating the innovation is greater than integrating it, the decision to lock on may be considered to be effective. It is preferable, at this moment, for innovation responsibility to be transferred from Pre-project to the Project Development.

We were looking at innovations in the car manufacturer's sense of the term, *i.e.* a technology or a service that does not exist within the company, although they might already be present at a competitor's or in another market sector.

The chosen approach takes in a global systemic standpoint of the touch-down process. For the purposes of the present article, we have generally used the term "process" in its broadest sense, *i.e.* including related organisation and instrumentation and decision process. Furthermore, process robustness is a relatively recent topic, notwithstanding Taguchi's work on robustness in general completed for many years (Taguchi and Clausing, 1990). We refer to the following definition to qualify process robustness:

> *'Capability to deliver expected results in the presence of unexpected adverse factors'.*
>
> (Chalupnik *et al.*, 2007)

In our case, this means delivering an optimised "touch-down" process that is fully able to withstand adverse factors, whether these are internal or outside the manufacturer's sphere.

17.1.2 Analysis of Five Vehicle Development Projects

The concept of "touching down" or integrating innovations in vehicle development projects was analysed from a vehicle project standpoint by tracing the appearance and disappearance of certain innovations throughout the vehicle design and development phases.

Our study is conducted on five recent Vehicle Development Projects: one of which was launched on the market at the beginning of 2010, one was discontinued and three are still under development. We traced the innovations flow from the Innovation

Seminar (held immediately prior to the formal Project start-date, "Intention" milestone, which aims to re-define and prioritise potential innovations to be integrated into vehicle projects) to the Contract milestone (which marks effective start of the development phase and is followed by industrialisation). See Figure 17.1. We chose the Vehicle Contract milestone as the reference milestone, because, from this stage, the vehicle content is frozen until the launch phase. The average time between the formal vehicle development start-date and the Vehicle Contract milestone is two years.

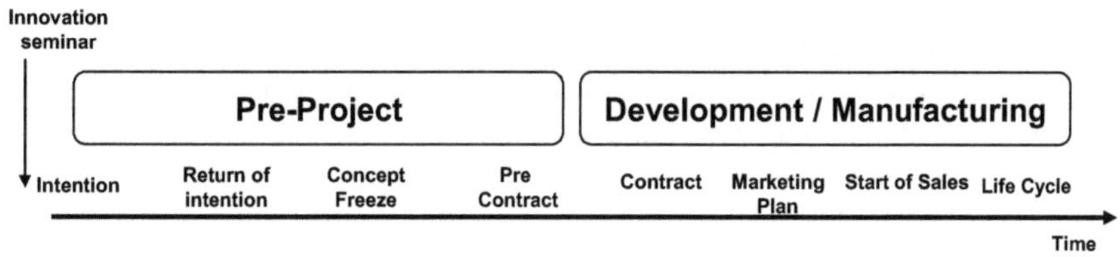

Figure 17.1. Vehicle Development Project, with milestones

A depth analysis based on 54 innovations targeted for three different vehicle projects defined at the end of the Innovation Seminar (after prioritisation) revealed that 37% of innovations find their way into the vehicle project by the Contract milestone stage. However, although some innovations are eliminated during the pre-project phase, others emerge as a result of market developments, customer expectations, competitors' activities and regulations, *etc.* In addition, some innovations are killed at a certain stage, only to be "resuscitated" later on.

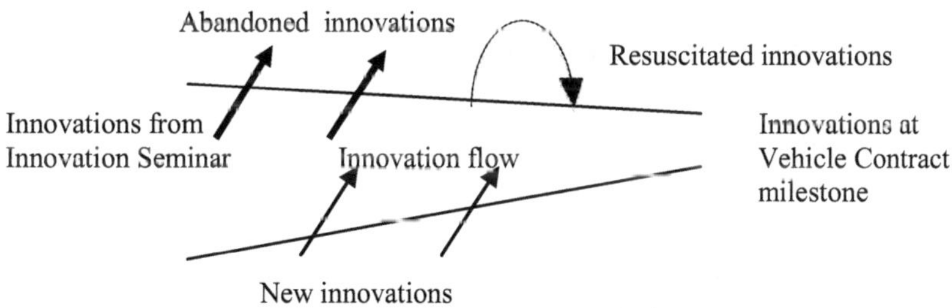

Figure 17.2. Innovation flow from Innovation Seminar to Vehicle Contract milestone

This analysis raises the pertinence of the actual innovation touch-down process: many innovations that ultimately land up in the vehicle development projects have not gone through the standard innovation process, consisting in identifying the innovations to be incorporated two years prior to the Contract milestone.

The factors that engender this situation may be internal, for instance:

- technical constraints of integrating certain innovations;
- economic evaluation of the innovation reveals no return on investment;
- a high degree of uncertainty (cost, volumes, potential customer value, reliability, *etc.*), which makes it difficult to make a decision at the right time;
- difficulty in developing the innovation in timescales compatible with vehicle development timescales;

- innovation maturity (Jahn and Binz, 2009). From a certain stage onwards, a low level of maturity may cause a lack of confidence in the innovation and hence a reticence to integrate it in vehicle project(s);
- lack of support from Top Management;
- inability to anticipate requirements upstream (changes in regulatory requirements, etc.).

Or external, for instance:

- sudden, unexpected changes in the market;
- abrupt changes in regulations (environmental, safety, *etc.*) making it necessary to discontinue certain innovations or integrate others;
- in the event of co-innovation (steadily increasing over recent years), partners' level of commitment.

This global analysis suggests the need to work on a more robust "touch-down" process. A robust process does not necessarily mean a lack of flexibility; indeed, a more flexible "touch-down" may enable innovations to be integrated at a later stage, if they are not too intrusive. In addition, the innovation development timescales may be very different, depending on their field of application (safety, performance, environment, passenger comfort). Some innovations may be developed over very short periods and be integrated at a late stage without any risk to vehicle project quality, if their intrusiveness is extremely low. They can be handled outside the vehicle project cycle.

17.2 Research Methodology

17.2.1 Introduction

Our methodology included six stages:

1. bibliographical research (exploratory rationale, inventory, options);
2. detailed diagnostic study, identification of initial solutions for progress;
3. modelling in order to gain a better understanding of the "touch-down" process;
4. descriptive analysis of three real cases of innovation "touch-downs", to enable comparison of the key points of the diagnostic and the modelling with actual cases;
5. formalisation of a "touch-down" theoretical model;
6. experimentation: specification logic for environmental innovations applied to future vehicle projects.

The work presented in the present article corresponds mainly to stages 1 to 4.

17.2.2 Results of Detailed Diagnostic Study

40 people were consulted, representing around 80 hours of interviews. Those consulted currently play or have in the past played a role in the "touch-down" process, *i.e.* they were new technology project managers, vehicle development project managers, financial project managers, technical managers (vehicle architecture, electronics, *etc.*), representatives of Pre-projects and of Marketing. High-level decision-makers were also questioned as part of our diagnostic study.

The aim was to gain a closer insight into real-life practices and at the same time to bring to light any differences between the official message and local perceptions.

The interviews were conducted in a semi-guided manner. The subjects covered were: innovation management by the motor manufacturer (as perceived by the people consulted), analysis of the current "touch-down" process for innovations integrated into vehicle projects (strong point advantages, areas for improvement), feedback on some specific cases as experienced by the people consulted.

In summary, the diagnostic study revealed four major points, outlined below, about which there were contradictory opinions:

- **Innovation requirement:** some of the people consulted were of the opinion that the requirement should be relatively broad, providing a mere framework for the innovation specification and not limiting the area of investigation. However, others believed that the requirement should be reduced to the bare necessities, thus limiting the area of investigation quite early on in the process in order to maximise its chances of success. It should be noted that not all innovations are derived from a formal innovation requirement; many arise spontaneously (proposals from the technical areas, R&D and from suppliers).
- **The "touch-down" process** must make reference to design/development standards, without being overly rigid. Excessive standardisation is perceived to be an obstacle to innovation, preventing good innovation opportunities from being taken up. The process should provide structure, but at the same time be flexible and able to be adapted to the different types of innovations - a single standard would not easily meet this requirement.
- **Evaluation and selection criteria**: the advantage of the criterion of cost-effectiveness (cost/value) that is widely used to decide whether or not an innovation should be assimilated into a vehicle is that it is consensual and easy to appropriate, particularly as the content and calculation methods for both costs and value criteria have recently been considerably optimised. However, cost-effectiveness reveals its limits in certain cases and sometimes needs to be completed by additional criteria relating to impact on image, contribution to corporate strategy, need to comply with regulations or to be in phase with the market and keep abreast of our competitors' market offerings.
- **Parties involved**: contradictions appear in particular with respect to the teams in charge of design and development of an innovation. For some of the people questioned, it is a good thing that these teams are not the same - they believe that it is necessary for team members working in upstream phases, where the creative input is high, to be different from those working downstream, where the final aim is very concrete (to industrialise the innovations). For others, the

weak point of the "touch-down" process is the split between the upstream and downstream phases - they feel that it is essential to maintain the same team (or some of the team-members) throughout the entire development process.

17.3 Examples of Innovations "Touch-down"

17.3.1 EasyDrive and EasyNav

Three instances of innovations "touch-down" were analysed regarding theoretical models (Lemoigne, 1990; Gidel and Zonghero, 2006) and these real cases were compared with the research hypotheses. In this way, we produced three "touch-down" pictures to provide an empirical description. We present in Table 17.1 two different cases: the first, EasyDrive is the result of a technology push and the second, EasyNav, was generated by market pull and was part of an "Open Innovation" process (Chesbrough, 2006).

Table 17.1. Comparison of the two cases

EasyDrive	EasyNav
Demand	
- Desire to set the manufacturer apart in the active safety field - An innovation that significantly changed key vehicle units (steering, chassis, suspension) gave rise to several research projects in the 1990s - These research projects, which to date had not been integrated in any vehicles, re-emerged within a vehicle development pre-project that was short on innovations.	- Initial opportunity came from the marketplace with the explosion of nomad navigation systems at very competitive prices, completely short-circuiting the automotive sector, which offered much more expensive integrated systems. - Without any requirement being expressed at the outset, an in-house technical team conversant with this technology grasped the opportunity to work on low-cost navigation solutions.
Integration ("touch-down") process	
- Development: innovation initiated by technical areas. From the outset, accent was laid on operational reliability with the aim of eradicating all potential risks. - Convergence: choice of technical solution. Production of prototypes to validate the solution and appropriation by key players. Discovery during the process ("Design Thinking" principle, Brown, 2008) that over and above active safety, driveability was revealed by the demonstrators to be a major aspect of differentiation. - Integration of the innovation into the vehicle project - the innovation was introduced at an early stage, because it had a high intrusiveness	- Development: the technical team made contact with several external companies specialised in nomad navigation systems. Intensive testing of the envisaged technical solutions was conducted in collaboration with the target companies, to identify potential solutions and to draft a requirement specification. - Convergence: production of very convincing demonstrators, which facilitated decision-making. - Integration of the innovation: consultation of all vehicle projects with the aim of integrating this innovation. From the start, the innovation was designed to be

level. The innovation was locked on to the vehicle project before the anticipated vehicle milestone. Excellent co-ordination between the different in-house entities and the key supplier. - The innovation was developed within the vehicle project: innovation development followed the set milestones, with good co-ordination between the parties involved (in-house entities and major suppliers). The quality of the innovation led to the manufacturer being awarded an "Innovation Trophy" in 2007.	transversal and was able to meet the technical constraints of different vehicle projects. - Development of the innovation within vehicle projects: no development in the strict sense of the term - it was a question of integrating a product into the vehicle projects, each of which had specific technical constraints that had previously been identified. This explains the extremely short development timescale compared to normal standards. It was possible to integrate the innovation at a very late stage because of its low level of intrusiveness; a more conventional back-up solution also existed.
Evaluation and decision criteria	
- Great uncertainty concerning the cost/value ratio of the innovation throughout the vehicle design and development phases (cost changes, changes in volumes, added value for the customer, *etc.*) which made it difficult to make a decision. Development of the innovation within a single vehicle project restricted potential volumes. - The normal evaluation criterion (cost/value) revealed its limits, due to very high initial investment (complexity of the innovation) and the low level certainty regarding sales assumptions.	- The cost/value evaluation was extremely advantageous for the manufacturer. Potential gains were proven (declared customer expectations, volumes, *etc.*). When the project was presented to Executive Management, the risk of not implementing the innovation was greater than the risk of implementing it. - Beyond the economic aspect, the first milestones of the innovation programme plan were all achieved and initial samples (initial parts from tooling) were convincing. - The innovation became a vital necessity in the economic context and was welcomed by the commercial network.
Roles played by the participants	
- Project bearers: the initial bearer had strong technical credentials and internal support. The development pilot also enjoyed in-house renown for his technical and managerial skills. - Sponsors: the innovation received support from Top Management from the outset, which was essential when there was major uncertainty concerning its profit-earning capacity. - "Touch-down" (convergence with Vehicle Project) network: an external consultant was sought during the pre-project phase to co-ordinate the different technical areas and to manage overall project progress towards convergence. Subsequent transfer between the pre-project and project phases went smoothly. Certain players in the pre-project phase continued to be involved until the industrialisation phase.	- Project bearers: a small, close-knit team from the start operating according to in-house entrepreneur principles (Bouchard, 2009). This team was convinced of the project's pertinence and its potential profit-earning capacity. The team itself "sold" the innovation directly to Vehicle Projects. - Sponsors: during the pre-project phase, the team benefit from a "Business Angel" which meant that the necessary budget was made available. When the innovation entered a more practical phase (demonstrators, business case creation, *etc.*), Top Management imposed the innovation on Vehicle Projects. - "Touch-down"(convergence with Vehicle Projects) network: this network included people who were extremely complementary, between the manufacturer's internal team

	which had worked on the project, representatives of the selected supplier who formed an integral part of the project and academic circles which had been consulted at the outset concerning work methodology. Furthermore, there was no "break" between the pre-project phase and the development phase; the team remained virtually the same.

17.3.2 Lessons Learned from the Cases Presented

These two cases illustrate very different innovation "touch-down" processes, particularly during the initial stages. The lessons learned from these two cases and from our diagnostic are the following:

- **Innovation reference standard:** the cases presented and the diagnostic demonstrate the need for a flexible and adaptable "touch-down" process. This could lead to proposals not for a single, unique "touch-down" process, but two or three processes as a function of the innovations being handled (intrusiveness, *etc.*).
- **Origin of the innovations**: it is necessary to strengthen the links between the technical areas, whose prime function is not innovation, but who may generate innovations by taking up opportunities, and the areas responsible for generating and managing innovation. Links are necessary for better communication between these two "worlds", via innovation-bearers and sponsors.
- **Moving the innovation downstream**: the two cases presented demonstrate the importance of a well-managed transfer between the pre-project phases and the development phase. All or some of the innovation project team members remained until industrialisation and launch. Without having to keep the same team from start to finish, it would appear to be necessary to keep at least one or two team members to provide project continuity and give a "running start" for successful convergence between the innovation and the target vehicle development project(s).
- **Innovation transversality**: innovation transversality, particularly if the innovation requires major investment, must be thought about in the very early stages. Technically, the constraints of rolling the innovation out to several vehicle development projects with, potentially, very different architectures must be taken into account from the outset. The cost of "re-contextualising" an innovation may be extremely high, sometimes requiring full system redesign. However, integrating all the technical constraints inherent in each target vehicle project can compromise innovation transversality. It is thus necessary to manage the tensions between this complexity, the impact of transversality on time-to-market and potential volumes.
- **Managing risk and projection into the future**: the innovation must be globally evaluated with respect to what it can contribute to the project and to the company in terms of profit-earning capacity, but also image, contribution to strategic objectives, *etc.* Sophisticated decision-making tools may help,

but they will never resolve the risk inherent in the decision to innovate, particularly as some criteria are difficult to quantify (impact on brand image, *etc.*). Moreover, improved projection into the future should allow learning curves for cost and volume assumptions to be defined further upstream in the process. Finally, from the vehicle development standpoint, it may be expedient to define the budget allocated to all (minor and ground-breaking) innovations from the start and to keep to this budget, despite the unexpected adverse factors that may emerge during the project.

- **Management support for innovations**: it appears to be essential to have the support of the "bearers" and "sponsors" (generally Top Management) of an innovation considered to be relevant by the company for it to succeed. These bearers and sponsors may support innovations during the decision stage by top management.
- **Co-ordination between technical areas and implementation of a "touch-down" network**: it is sometimes necessary to call upon a neutral party, or even an outside consultant to provide co-ordination between the technical areas. Moreover, it is essential to form a dedicated network of people to ensure successful final "touch-down", *i.e.* convergence with the vehicle project. It should be organised, structured and "open" - all potential resources should be used to facilitate convergence, be these internal or external (OEMs, academic circles, *etc.*).

17.4 Discussion: Interpreting the Research

17.4.1 Potential Improvements to the Innovation "Touch-down" Process

Ideas to improve the way in which innovations are assimilated into Vehicle Development Projects were formulated following the extremely detailed diagnostic study (40 interviews) and were developed after the work carried out on real innovation cases and analysed using a reference standard model filter. At this stage, the hypotheses expressed should not be considered to be recommendations for a more robust convergence process - they are factual observations based on a large number of interviews, bibliographical research and real-case analysis. They will undergo additional validation, notably during the specification phase - an experimental protocol will be defined to test the pertinence of these hypotheses on innovation projects undergoing final "touch-down", *i.e.* convergence with a vehicle project.

In summary, in order to improve robustness of innovation convergence with Vehicle Development Projects, we have marked out four fundamental hypotheses:

- **Need to structure requirement formalisation** according to a partnership rationale, involving both innovation players and vehicle project players. It is necessary to accurately define the activities expected by Vehicle Projects and to identify the innovations that allow these goals to be achieved. This clarification must enable the people responsible for designing the innovations to appropriate the vehicle projects' expectations by limiting the area of investigation to the bare minimum. The objective is to define a simple strategy that is carried out and adjusted by means of regular feedback between the innovation project players and the vehicle project players. Methods such as Quality Function Deployment (Akao, 1990 ; Shiba *et al.*, 1993) may provide a better understanding and deployment of the requirement.

- **Need for one or more generic "touch-down" processes**, selected as a function of the specific type of innovation (level of intrusiveness in the vehicle project, need to respond swiftly to market demands, *etc.*) and taking into account the key events of touch-down (firing slot). These processes would form a framework that should be adaptable to each specific case.
- **Need to facilitate decision-making,** based on information from appropriate evaluation and decision-making criteria related to the concerned field (example: safety, passenger comfort, environment/CO2) and manage risk by working on different scenarios relating to volumes, costs, income and profit-earning capacity. It is essential to have, at the opportune moments (when arbitration/decisions have to be made) pertinent criteria and data (including the associated level of risk), by sufficiently anticipating innovation project design/development constraints - principle of irreversibility (Giard and Midler, 1993; Chvidechenko and Chevalier, 1997). This should help towards more robust decisions being made (Ullman, 2001).
- **Need to identify the players and define their role** ("project bearer", "sponsors", "touch-down network", focusing effort - "running start" to enable convergence of all players involved. A key component is the complementary disciplines of the people forming part of the "touch-down" network.

17.4.2 First Approach to Establishing an Innovation "Touch-down" Model

At present, there is an identical "touch-down" process for integration of all innovations within Vehicle Projects, no matter what type. However, innovations are, by their very nature, unique in character, which could result in defining numerous "touch-down" processes. This would run counter to the need for standardisation.

Between these two extremes, it appears to us to be necessary to have two or three "touch-down" processes and not a single process, to be applied depending on the type of innovation (A, B or C). Figure 17.3 illustrates this proposal.

This outline model constitutes an initial idea for further development. This, of course, presupposes that classes of innovations are defined according to a multi-criteria approach.

Abundant literature exists on the methods of classifying innovations in relation to the specific objective: impact on integration with vehicle projects (Garcia and Calantone, 2002; Velloso-Rodriguez and Truchot, 2009).

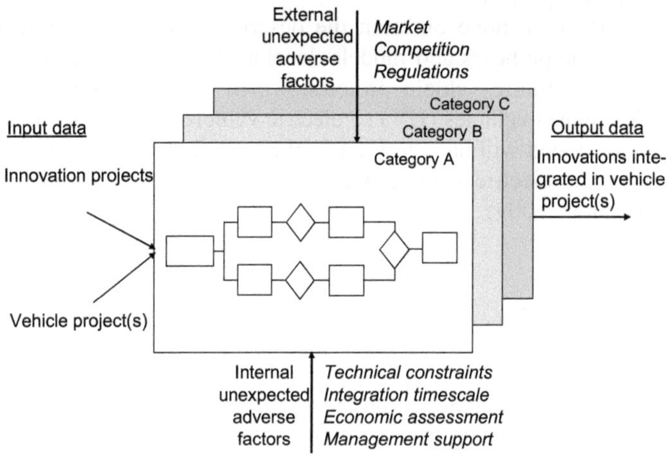

Figure 17.3. "Touch-down" process as a function of the type of innovation

17.5 Conclusion and Perspectives

The present article identifies the essential elements that typify the issue of "touch-down"; it identifies several concepts ("lock-on", "touch-down network", *etc.*), which require further in-depth study. The purpose of this work is to help to increase the number of innovations with high added value for the customer that find their way into future vehicles.

Our detailed diagnostic study and analysis of three real innovation cases has brought to light some major factors which could have an impact on "touch-down" robustness - the need for a formal structured requirement for the people dealing with innovation projects and those in charge of vehicle development projects, an adaptive process or processes depending on the type of innovation, definition of appropriated evaluation/decision-making criteria for the concerned people plus management of uncertainty risk, and the need for key players ("innovation bearers", "sponsors") to "buy into" the innovation in order to facilitate their final implementation. These different variables enable us to act on the robustness of the touch down process, not from a statistical point of view, but from a process, an organisational standpoint.

Application of these hypotheses for improvement should help to make integration of innovations into the manufacturer's future vehicle development projects. However, this research programme is only at the conception/description stage. All the factors driving robustness will have to be detailed and validated. The experimental protocol and the "touch-down" model(s) that will emerge will embrace all the lessons drawn from our diagnostic and from our real-case investigation. Following an experimental phase, these lessons may result in a list of recommendations to ensure successful touch-down, the aim being to propose one or more very specific models to in-house clients working on innovation projects or vehicle development projects.

Then, the aim is to move on from the descriptive stage to formulation of a specification - the hypotheses and models developed will be tested for integration of innovations into future vehicle projects, which may be internal combustion Vehicles (ICE), hybrid vehicles (HV) or electric vehicles (EV). The "touch-down" models to be developed will have to include the specific features of these markets, in particular for the electric vehicle, which introduces a new development logic (Midler and Beaume, 2009).

17.6 References

Akao Y (1990) Quality function deployment - integrating customer requirements into product design. Productivity Press, Cambridge, MA, US

Bouchard V (2009) Intrapreneuriat - innovation et croissance: Entreprendre dans l'entreprise. Dunod, France

Brown T (2008) Design thinking, Harvard Business Review, 86(5): 84-92

Buet G, Gidel T, Millet T (2008) New technologies "touch down process". In: Proceedings of the International Conference on Integrated Design and Manufacturing in Mechanical Engineering (IDMME 2008), Beijing, China

Chalupnik MJ, Wynn DC, Eckert CM, Clarkson PJ (2007) Understanding design process robustness: A modelling approach. In: Proceedings of the 16th International Conference on Engineering Design (ICED'07), Paris, France, pp 455-456

Chesbrough H (2006) Open innovation: The new imperative for creating and profiting from technology. Harvard Business School Press, Boston, MA, US

Chvidchenko I, Chevalier J (1997) Conduite et gestion de projets. Cédapuès Editions, Paris, France

Garcia R, Calantone R (2002) A critical look at technological innovation typology and innovativeness terminology: A literature review. The Journal of Product Innovation Management, 19: 110-132

Giard V, Midler C (1993) Pilotages de projet et enterprises. Economica, Paris, France

Gidel T, Zonghero W (2006) Management de projet 2: Approfondissements. Hermes Science Publications, Paris, France

Jahn T, Binz H (2009) A highly flexible project maturity management method for the early phase of product development. In: Proceedings of the 17th International Conference on Engineering Design (ICED'09), Stanford, CA, US

Lemoigne JL (1990) La modélisation des systèmes complexes. Dunod, France

Midler C, Beaume R (2009) Project-based learning patterns for dominant design renewal: The case of electric vehicle. International Journal of Project Management, 28(2): 142-150

Shiba S, Graham A, Walden D (1993) A new American TQM - four practical revolutions in management. Productivity Press, Cambridge, MA, US

Taguchi G, Clausing D (1990) Robust quality. Harvard Business Review, 68(1): 65-75

Ullman DG (2001) Robust decision-making for engineering design. Journal of Engineering Design, 12(1): 3-13

Velloso-Rodriguez K, Truchot P (2009) Classification des projets d'innovation d'après leur but d'origine. Confere'09, Marrakech, Maroc

Index of Contributors

MIX
Papier aus verantwortungsvollen Quellen
Paper from responsible sources
FSC® C105338

If you have any concerns about our products,
you can contact us on
ProductSafety@springernature.com

In case Publisher is established outside the EU,
the EU authorized representative is:
Springer Nature Customer Service Center GmbH
Europaplatz 3, 69115 Heidelberg, Germany

Printed by Libri Plureos GmbH
in Hamburg, Germany